HeadStar

Primary

Maths Problem Solving, Reasoning & Investigating

Year 5

Lizzie Marsland
Susannah Palmer

Acknowledgements:

Author: Lizzie Marsland, Susannah Palmer

Series Editor: Peter Sumner

Cover and Page Design: Kathryn Webster, Jo Sullivan

The right of Lizzie Marsland and Susannah Palmer to be identified as the authors of this publication has been asserted by them in accordance with the Copyright, Designs and Patents Act 1988.

HeadStart Primary Ltd
Elker Lane
Clitheroe
BB7 9HZ

T. 01200 423405
E. info@headstartprimary.com
www.headstartprimary.com

Published by HeadStart Primary Ltd 2019 © **HeadStart Primary Ltd 2019**

A record for this book is available from the British Library -
ISBN: 978-1-908767-61-5

INTRODUCTION

<u>Year 5: NUMBER - Number and place value</u>

CONTENTS

Year 5: NUMBER - Addition and subtraction

Year 5: NUMBER - Multiplication and division

CONTENTS

Year 5: NUMBER Fractions (including decimals and percentages)

CONTENTS

Year 5: MEASUREMENT

CONTENTS

INTRODUCTION

These problems have been written in line with the objectives from the Mathematics Curriculum. Questions have been written to match all appropriate objectives from each content domain of the curriculum.

Solving problems and mathematical reasoning in context are difficult skills for children to master; a real-life, written problem is an abstract concept and children need opportunities to practise and consolidate their problem solving techniques.

As each content domain is taught, the skills learnt can be applied to the relevant problems. This means that a particular objective can be reinforced and problem solving and reasoning skills further developed. The first section of each content domain is intended to provide opportunities for children to practise and consolidate their problem solving skills. Each page has an identified objective from the National Curriculum; the difficulty level of the questions increases towards the bottom of each page, thus providing built-in differentiation.

Mastering a skill involves obtaining a greater level of understanding of the skill, the ability to transfer and apply knowledge in different contexts and explaining understanding to others.

The MASTERING and INVESTIGATION sections provide extra challenges as children's problem solving skills and confidence increase. The problems in the MASTERING sections encompass several objectives from the relevant curriculum domain. The INVESTIGATION covers objectives from across the whole curriculum.

At HeadStart, we realise that children may need more space to record their answers, working out or explanations. It is recommended, therefore, that teachers use their discretion as to where children complete their work.

Since a structured approach to problem solving supports learning, developing a whole-school approach is highly recommended.

Throughout this book, 6 children are solving problems.

Their names are:

Lola

Tom

Ayesha

Charlie

Sadia

Bilal

NUMBER

Number and place value

These are all about number and place value!

27-9-23

Read and write numbers to at least 1,000,000

1   Mrs Holly asks her class to write the number **270** in words. How should the class write this number?

two hundred and seventy

2   Lola writes the number **eight hundred and fifty two** in numerals. What does Lola write?

8 52

3   Mr Ahmed is writing a sign in his shop. A television costs **one thousand, two hundred and ninety nine pounds**. How should Mr Ahmed write the cost of the television in numerals?

£1,299

4   Charlie writes the number **eighteen thousand and sixty two** in numerals. What does he write?

18 062

5   Bilal writes the number **10,287** in words. What does he write?

ten thousand two hundred and eighty seven

6   Miss Smith has **twenty two thousand, eight hundred and sixty five pounds** in her bank account. Write how much money she has in numerals.

£22,865

7   In a test, Tom has to write the number **567,874** in words. What should he write? five hundred and sixty seven thousand, eight hundred and seventy four

8   Mrs Harris wins **£1,673,202** on the lucky lottery. Write this figure in words.

one million, six hundred and seventy three thousand two hundred and two pounds

Read and write numbers to at least 1,000,000

1 Charlie is writing a birthday card for his grandad, who is going to be **78** years old. How should he write this number in the card using words?

seventy-eight

2 A farmer has **two hundred and thirty five** chickens. How should he write this as a numeral?

_235_

3 Miss Heaton asked Bilal to write the number **893** on the whiteboard in words. How should he write the number?

eight hundred and ninety-three

4 Tom reads the number **two thousand, five hundred and twenty four**. How should he write this number in numerals?

_2524_

5 Ayesha's mum writes a cheque for **£7836** to pay for her car. How should she write this number in words?

seven thousand eight hundred and thirty six _pounds_

6 Lola's mum has forgotten how to write the number **230,987** in words. Lola shows her. What does she write? _two hundred and thirty thousand nine hundred and eighty-seven_

7 Charlie reads the number **887,342**. He says this number is written as **"Eight thousand and seventy eight, three hundred and forty two."** Is Charlie correct? Explain your answer. _No because he turned_ _8(87)342 into 8(78)342_

8 Write the number **one million, one hundred and sixty three thousand, four hundred and fifty seven** in numerals.

checked

_1,163,457_

2

Name ....

Order and compare numbers to at least 1,000,000

1   Put the numbers below in order of size from smallest to largest.

        (3)   (2)   (1)   (4)

        5652   **4652**   **3652**   **9652**

3652, 4652, 5652, 9652

2   Bilal is trying to decide which number is larger: **22,292** or **22,392**. Which one should he choose?

22 392

3   Ayesha puts these numbers in order of size from largest to smallest. What should her list look like?

        352,200   **352,267**   **352,298**

352,298   352,267   352,200

4   Mrs Bell asked her class to write down the number which is larger than **632,223** but smaller than **632,225**. What number should they write?

632 224

5   Which number is smaller: **763,892** or **736,982**?

736 982

6   Which number is larger: **999,998** or **999,989**?

999998

7   Sadia wants to put these numbers in order of size from smallest to largest. What should her list of numbers look like?

corrected

        989,000   **998,000**   **1,900,000**   **1,000,009**

989,000, 998,000, 1000009, 1900000

8   Write a number larger than **546,351** but smaller than **546,357**.

546352, 546353, 546354, 546355

27-9-23

Order and compare numbers to at least 1,000,000

1 Tom had to find out which number was smaller: **1782** or **1972**. Which one should he choose?

1782

2 Ayesha put the following numbers in order from largest to smallest. What should her new list look like?

| 3010 | 3005 | 3000 | 3205 |

3205, 3010, 3005, 3000

3 Tom's teacher asks him to write a **four-digit** number which is larger than **5435**, but less than **5437**. Which number should he write?

5436

4 Put these numbers in order from largest to smallest.

| 22,532 | 22,893 | 22,212 | 22,793 |

22,893, 22,793, 22,532, 22,212

5 Write down the number that is smaller than **432,354** and larger than **432,352**.

432,353

6 Sadia is trying to put these numbers in order from smallest to largest. What should her answer be?

| 642,873 | 783,922 | 642,837 | 738,299 |

642,837, 642,873, 738,299, 783,922

7 Which number is larger: **1,982,998** or **1,982,989**?

checked 27-9-23

1,982,998

Determine the value of each digit in numbers up to 1,000,000

1 There are **252** pupils at Telfroy Primary School. How many **hundreds** are there?

2 Lola counts **3325** flowers in her garden. How many **thousands** are there?

3 Bilal partitions the number **9728** into **thousands**, **hundreds**, **tens** and **ones**. Which digit represents the **hundreds**?

4 Mr Wright asks his class to partition **18,364** into **thousands**, **hundreds**, **tens** and **ones**. Which digit represents the **ones**?

5 Charlie has to write down the value of the digit **6** in the number **48,263**. What should he write?

6 Would you rather have the value of the digit **eight** in **£70,892** or in **£78,922**? Explain your answer.

7 Which is more: the value of the **7** in **967,899** or the value of the **7** in **456,745**?

8 What is the value of the digit **9** in the number **942,864**?

Name ....

Determine the value of each digit in numbers up to 1,000,000

1 Lola partitions **482** into **hundreds**, **tens** and **ones**. Which digit represents the **tens**?

2 What is the value of the digit **7** in the number **2793**?

3 Sadia partitions the number **18,264**. What is the value of the digit **eight**?

4 Ayesha's dad has forgotten about place value. He tells Ayesha that the digit **4** in the number **23,649** is worth **4000**. Is he correct? Explain your answer.

5 Charlie says, "The value of the digit **5** in the number **562** is **50**." Is he correct? Explain your answer.

6 Mrs Marshall writes down the number **349,245** on the whiteboard. She asks the class what the value of the digit **nine** is. What should the answer be?

7 Would you rather have the value of the digit **9** in the number **£843,933** or **£891,567**? Explain your answer.

8 What is the value of the digit **5** in the number **1,549,344**?

6

Name ....

Count forwards in steps of powers of 10 (100 or 1000)

1 Sadia has **78** shiny stickers. Her mum buys her **100** more. How many does she have now?

2 There are **142** people on the train. **One hundred** more people get on at Huston. How many are on the train now?

3 Count forwards **1000** from **2342**. What number do you get to?

4 Start at **456** and count on **two** more **hundreds**. What number do you count to?

5 At a football match, there are **5682** Hilly fans. There are **100** more Kinley fans. How many Kinley fans are there?

6 Lola starts at **2652** and counts on **two** more **hundreds**. She counts to **2952**. Is she correct? Explain your answer.

7 Count forwards **4000** starting at **70,246**. What number do you get to?

8 Bilal starts at **68,753**. He counts forwards **five thousand**. He then counts on **two** more **hundreds**. What number does he count to?

7

Name ....

Count forwards in steps of powers of 10 (10,000 or 100,000)

1 There are **12,900** ants in Nest A and **10,000** more in Nest B. How many ants are there in nest B?

2 A shop sells **25,232** sweets in a year. It sells **10,000** more chocolates. How many chocolates does the shop sell?

3 A silver sports car costs **£15,600**. A red sports car costs **£10,000** more. How much does the red sports car cost?

4 Ayesha counts forward **100,000** more than **200,800**. What number does she count to?

5 In the city library, there are **673,892** fiction books. There are **100,000** more non-fiction books. How many non-fiction books are in the library?

6 Start at **73,892** and count on **ten thousand**. What number do you count to?

7 Count forwards **400,000** starting at **348,946**. What number do you count to?

8 Sadia starts at **248,651**. She counts forward **10,000**. She then counts on **two** more **hundred thousands**. What number does she count to?

Count backwards in steps of powers of 10 (100 or 1000)

1 The supermarket has **2830** bananas. They sell **1000** on Saturday. How many bananas are left?

2 There are **5982** fish in the lake. The fishermen catch **1000**. How many fish are left in the lake?

3 Tom counts back **100** from **8623**. What number does he count to?

4 Charlie's big brother has saved **£7843**. He spends **£1000** on a holiday. How much money does he have left?

5 At the pop concert there are **12,753** people. **1000** people leave early. How many people are left?

6 At a football match there are **15,576** Melchester fans. There are **two thousand** less Welsea fans. How many Welsea fans are there?

7 Ayesha starts at **53,293**. She counts backwards to the number **49,293**. How many **thousands** has she counted back?

8 Start at **83,742** and count back **five thousands**. Then count back **six hundreds**. What number do you reach?

Count backwards in steps of powers of 10 (10,000 or 100,000)

1 There are **22,320** bees making honey in the hive. **10,000** fly off to visit flowers. How many bees are left in the hive?

2 Count back **10,000** from the number **182,755**. What number do you reach?

3 What number do you reach when you count back **100,000** from **426,342**?

4 Park A has **354,897** flowers. Park B has **100,000** less. How many flowers are in Park B?

5 A supermarket sold **325,900** apples. This is **100,000** more than last year. How many apples did the supermarket sell last year?

6 Count back **10,000** from **67,983**. What number do you reach?

7 Strikemore football team has **£873,920**. They spend **£100,000** on a new player and **£100,000** on a new pitch. How much money do they have left?

8 Bilal starts at **983,892**. He counts back **40,000**. He then counts back **600,000**. What number does he count to?

28-9-23

Count forwards or backwards in steps of powers of 10 (mixed)

1 Start at **8321** and count forwards **one hundred**. What number do you count to?

_8421_

2 Count back **10,000** from the number **72,800**. What number do you count to?

_62 800_

3 Mrs Shaw has **£12,874** in her bank account. She spends **£1000** on a new television. She then spends **£1000** on new clothes. How much money does she have left?

_£10 874_

4 At the cricket match, there are **22,753** fans. **1000** fans leave early. How many fans are left?

_21 753_

5 Lola counts forwards **30,000** from **58,345**. What number does she reach?

_88 345_

6 Tom writes down the number **76,847**. He counts forward **five thousands**. What number does he count to?

_81 847_

7 Charlie starts at **896,475**. He counts backwards to the number **396,475**. How many **hundred thousands** has he counted back?

_5_

8 Start at **987,432**. Count back **three hundred thousand**. Then count forward **two hundreds**. What number do you reach?

_687 632_

Count forwards or backwards in steps of powers of 10 (mixed)

1 There are **5892** trees in the forest. **1000** are chopped down. How many trees are left in the forest?

4892

2 Count forwards **100** from the number **5755**. What number do you reach?

5855

3 This year **19,875** people visited the museum. This is **1000** more than last year. How many people visited the museum last year?

18875

4 Mr Taylor writes down the number **352,834**. He asks his class to count back **100,000**. What number should they say?

252834

5 What number do you reach when you count back **10,000** from **68,342**?

58342

6 Tom writes down the number **76,847**. He counts forwards **five thousands**. He then counts on **100**. What number does he count to?

81947

7 Lola starts at **283,253**. She counts backwards to the number **223,253**. How many **ten thousands** has she counted back?

6

8 Start at **498,534** and count on **three** more **hundred thousands**. Then count back **four tens**. What number do you reach?

checked

500494 798494

NUMBER - Number and place value

Interpret negative numbers in context

1 The thermometer is at **−2°C**. What is **three degrees** colder?

$-5°C$

2 The temperature in Portugal is **9°C**. The temperature in England is **10°C** colder. What is the temperature in England?

$-1°C$

3 The temperature in Antarctica is **−16°C**. The temperature drops by **2 degrees**. What is the temperature now?

$-18°C$

4 What is **5** less than **−7**?

-12

5 What is the difference between **3** and **−6**?

9

6 A freezer's temperature is **−18°C**. Lola leaves the freezer door open and the temperature rises by **4 degrees**. What is the temperature now?

$-14°C$

7 Bilal says, "**9** more than **−3** is **12**." Is he correct? Explain your answer.

No as it is -3 not 3

8 Tom starts at **−9**. He counts back **5**. He then counts forward **8**. What number does he count to?

-6

Name ....

NUMBER - Number and place value

Year 5

Count forwards and backwards with positive and negative whole numbers through zero

1 Ayesha is counting backwards. She starts at **4** and counts back **5**. What number does she count to?

-1

2 Miss Small asks Class 5 to count back **4** from **−2**. What number should they say?

-6

3 Which number would Bilal say if he counted **9** forwards from **−6**?

-15

4 In Spain, the temperature is **15°C**. In the United Kingdom, the temperature is **18 degrees** colder. What is the temperature in the United Kingdom?

-3°C

5 On a negative number line, Sadia counts back **11** from **−3**. What number does she count to?

-14

6 The temperature at the North Pole on Monday was **−16°C**. On Tuesday, the temperature was **6 degrees** colder. What was the temperature on Tuesday?

-22°C

7 Tom has **£9** in his bank. He spends **£15** in the shop, using his debit card. How much does his bank account show? (His bank account can show minus numbers.)

£-6

8 Charlie starts at **−2** on the negative number line. He counts back **25**. He then counts forwards **3**. What number does he reach?

-24

Round any number up to 1,000,000 to the nearest 10 and 100

1 Bilal pours **1890 millilitres** of apple juice into a jug. Is this nearer to **1800** or **1900 millilitres**?

2 Charlie rounds **2387** to the nearest **100**. What number should he round to?

3 Mrs Barnes rounds **12,121** to the nearest **ten** on the whiteboard. What number should she write?

4 Miss Lamb's new sofa cost **£5285**. What is the cost of this to the nearest **£100**?

5 Mrs Jones has saved **£10,422** in her bank account. How much money has she saved to the nearest **£10**?

6 A car costs **£12,389**. How much does the car cost to the nearest **hundred pounds**?

7 Round the number **452,834** to the nearest **hundred**.

8 Tom rounds the number **873,283** to the nearest **hundred**. He writes down **874,300**. Is he correct? Explain your answer.

Round any number up to 1,000,000 to the nearest 10 and 100

1 The sandwich shop sold **112** cheese sandwiches. How many sandwiches did the shop sell to the nearest **100**?

2 Round the number **89** to the nearest **ten**.

3 Tom runs **2289 metres** around the park. Is this nearer to **2000** or **3000 metres**?

4 Mr Jackson asks his class to round **1439** to the nearest **hundred**. What number should they say?

5 Sadia rounds the number **82,382** to the nearest **hundred**. What number should she say?

6 Lola rounds **938,243** to the nearest **100**. Her answer is **938,240**. Is she correct? Explain your answer.

7 Round the number **1,394,506** to the nearest **10**.

8 At a football match, there are **4876** Redpool fans. There are **4212** Sandchester United fans. How many fans are there altogether to the nearest **hundred**?

Round any number up to 1,000,000 to the nearest 1000, 10,000 and 100,000

1 At Spring School, there are **2245** pupils. Round this number to the nearest **thousand**.

2 There are **6,388** Brunlee fans at the football match. Round this number to the nearest **thousand**.

3 Mr Potter buys a new car for **£13,342**. Is this nearer to **£14,000** or **£13,000**?

4 Sadia rounds **52,743** to the nearest **10,000**. What number does she say?

5 Ayesha rounds the number **389,000** to the nearest **100,000**. What should her answer be?

6 Round the number **762,340** to the nearest **ten thousand**.

7 In the park, there are **997,382** flowers. Is this nearer to **900,000** or **1,000,000**?

8 Mr Davies has **£124,679** in his bank account. He takes out **£2000**. Estimate how much money he has left to spend, to the nearest **ten thousand**.

Name ....

Round any number up to 1,000,000 to the nearest 1000, 10,000 and 100,000

1 Tom has **8937** stickers. How many stickers does he have to the nearest **1000**?

2 Which number below is the answer to **83,420** rounded to the nearest **1000**?

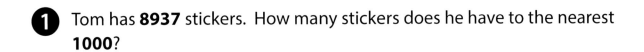

83,000   **83,400**   **84,000**

3 Big Ben is about **9600 centimetres** tall. What is the height to the nearest **1000 centimetres**?

4 Bilal rounds the number **6592** to the nearest **thousand**. What should his answer be?

5 Mrs Bubble asks Class 5 to round the number **234,322** to the nearest **hundred thousand**. What number should they say?

6 Charlie says that **33,678** rounded to the nearest **thousand** is **33,600**. Is he correct? Explain your answer.

7 Round **783,844** to the nearest **10,000**.

8 Sadia rounds the number **1,389,236** to nearest **100,000**. What should her answer be?

Read Roman numerals to 1000 (M) and recognise years written in Roman numerals

1 Ayesha buys **XII** eggs at the market. How many eggs does she buy?

12 ✓

2 Emperor Nero fiddled while Rome burnt in **LXIV** AD. What year did this event happen?

64 AD

3 Spartacus, the gladiator, has won **LXXX** fights. How many fights has Spartacus won?

80 ✓

4 The Roman army marched **CCLXXX** miles to Rome. How many miles did they march?

Checked

280 ✓

5 Augustus was born in the year **CM** BC. Which year was Augustus born?

900 BC ✓

6 An inscription states that Rome was founded in **DCCLIII** BC. Which year was Rome founded?

753 BC

7 There were **DCCLXX** Romans relaxing in the Roman baths. How many Romans were in the baths?

✓ 770

8 Write the year you were born in Roman numerals.

MMXIV ✓

Read Roman numerals to 1000 (M) and recognise years written in Roman numerals

1 There were **XXX** legions in the Roman army. How many legions were there?

30 ✓

2 Hadrian's wall was built **LXXIII** miles long, to separate the Romans from the Picts. How long was the wall in miles?

73 ✓

3 Julius Caesar became ruler of Rome in **XLIV** BC. Which year was this?

44BC ✓

4 Tarquinius, the gladiator, fought in the Colosseum in **CCLII** BC. Which year was this?

252BC ✓   checked

5 Julia and Marcus got married in the year **CDV** BC. Which year was this?

405BC ✓

6 Maximillian marched his army **CXIV** miles to battle. How many miles did they march?

114 ✓

7 Atticus inscribed the year **CDXIII** onto a rock. What number did he inscribe?

483   413

8 Write the year you are in now in Roman numerals.

MMXXIII ✓

W 19   20-1-24

Solve mixed number problems

1 Mr Bishra wants to write **six hundred and twenty thousand, five hundred and twenty one** as a numeral. What should he write?

6~~25221~~   620 521 ✓

2 In the number **364**, what does the digit **6** represent?

60 / six tens ✓

3 The temperature in Moscow was **–2°C**. The temperature dropped by **2 degrees**. What was the temperature then?

~~-4° 0°C~~ (-4°C) ~~-4 -3 -2 -1 0~~

4 Julius was born in the year **CCCVI** AD. Which year was he born?

306 ?

5 Sadia thought of a number that was smaller than **39,829** but bigger than **39,827**. What was her number?

39828 ✓

6 Ayesha rounded **872,278** to the nearest **10,000**. What was her answer?

~~870 0000~~   870,000

7 Write the number **1,873,233** in words. One million, eight hundred and seventy three thousand, two hundred and thirty-three

8 The temperature was **–8°C**. It rose by **11°C**. What was the new temperature?

3°C   3°C

checked

20-1-24 W19   3.10.23

Solve mixed number problems

1 Ayesha writes down the number **763**. What is the value of the digit **3**?

3 ones

2 Tom starts at **4** and counts back **8**. What negative number does he count to?

_-4_

3 Which number is bigger: **87,482** or **87,428**?

_87,482_

4 Round the number **894,389** to the nearest **thousand**.

_894000_

5 What does the **8** represent in the number **983,422**?

_80,000_

6 Write the number **six hundred and forty three thousand, two hundred and eight** in digits.

_643208_

corrected
checked

7 The temperature in Paris is **1°C**. It is **4°C** colder in Manchester. What is the temperature in Manchester?

_-3°C_

8 Lola writes the number **1,372,495** in words. What does she write?

one million, thirty three hundred and seventy two thousand four hundred and ninty-five

W 39
9-6-24
checked

Solve mixed number problems

1 Tom has **1036** animal stickers. He collects **1000** more. How many stickers does he have altogether?

2036

2 Sadia counts back **10,000** from **24,892**. What number does she count to?

14892

3 Lola writes the number **83,922**. What is the value of the digit **9**?

900

4 Cato joined the Roman army in the year **LXXII** AD. What year did he join the army?

72 AD

5 Which negative number would Ayesha say if she counted **7** back from **3**?

-4

6 Lola says that **12,578** rounded to the nearest **thousand** is **12,000**. Is she correct? Explain your answer. No as the digit 5 needs to be rounded up

7 Write the number **832,399** in word form. eight hundred thirty-two thousand three hundred and ninety-nine.

8 Tom is trying to decide which number is bigger: **1,543,234** or **1,458,999**. Which number should he choose?

W 39

9-6-24

checked

Solve mixed number problems

1 There were **32,456** people at a rugby match. Round this number to the nearest **ten thousand**.

30,000

2 Ayesha had to write the value of the digit **9** in the number **2,953,786**. What should her answer have been?

900 000

3 Write the number **thirty seven thousand, six hundred and twenty four** in numerals.

37,624

4 Tom had to round **3556** to the nearest **100**. What was his answer?

3600

5 What is **4,765,728** rounded to the nearest **10**?

4765730

6 Write these **four** numbers in ascending order (from smallest to largest):

1   4   3   2

9253   **9835**   **9753**   **9393**

7 Sadia wanted to know the value of the digit **2** in the number **1,244,516**. What is the answer?

200 000

8 Write the number **four million**, **seven hundred and sixty five thousand, two hundred and twelve** in numerals.

4,765,212

24

Name ....

MASTERING

Number and place value

You'll need your full powers of concentration for these!

1　Write the following numbers in descending order:

465,465　　　**654,465**　　　**456,465**　　　**546,465**　　　**645,465**

2　Tom has climbed **900 m** from the car park to the top of a hill.
The temperature drops **1°** every **150 m** he climbs up the hill.
At the top of the hill the temperature is **−2°**.

How many metres above the car park should the temperature be **2°**?

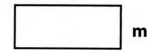 m

---

3　Ayesha and Lola are staying on the **12th** floor of a hotel.
They want to go swimming.
The pool is in the basement.
Ayesha gets the lift and presses the button **−1**.
Lola races Ayesha down the stairs.

How many flights of stairs does Lola have to run down?

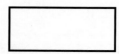

　　　25　　　Name ....

4　**a**　Charlie thinks of a number. When rounded to the nearest **10,000** it is **540,000**.

Bilal says, "Your number must be between **540,000** and **544,999**." Is Bilal correct?

| Yes / No |

Explain your answer.

....

....

b　Tom's number is a multiple of **9**.

The **tens** of **thousands** digit, the **thousands** digit and the **hundreds** digits are consecutive multiples of **3** written in ascending order.

The **tens** digit is a **third** of the value of the **thousands** digit.
What is Tom's number?

| |

5　**a**　What year is it now? Write it in Roman numerals.

| |

b　Write your birthday in Roman numerals.

| |

c　Find the difference between the **two** dates and write your answer in Roman numerals.

| |

Name ....

6　**a**　Write these numbers in ascending order:

880,808　　　808,880　　　800,888　　　880,880　　　880,088

b　Which number is the smallest?
Explain how you know

....

....

7　Write **six 6-digit** numbers where the digit sum is **9** and the **tens of thousands** digit is **5**.

a　What is the smallest number?

b　What is the largest number?

c　What is the difference between the largest and smallest number?

8 Charlie wrote a number and then rounded it to the nearest **thousand**.
His answer was **54,000**.
Sadia wrote a different number and rounded it to the nearest thousand.
Her answer was also **54,000**.
What is the largest possible difference between Charlie and Sadia's numbers?

9 Ayesha has a bag of tiles numbered **0-9**.
She uses all the tiles to make the largest even **5-digit** number and the smallest odd **5-digit** number possible.
She can use each tile only once.
What is the difference between these **two** numbers?

10 Lola recorded the temperature every **two** hours for **12** hours while on holiday. Here are her results:

Time	08:00	10:00	12:00	14:00	16:00	18:00	20:00
Temperature	−2°C	−1°C	2°C	4°C	1°C	−3°C	−2°C

a Between which **two** times was the greatest drop in temperature?

[ ] **and** [ ]

b Between which **two** times was the greatest rise in temperature?

[ ] **and** [ ]

c Do you think Lola was on holiday in Summer or Winter?
Explain your answer.

....

....

11 Here is part of a number line. Fill in the missing numbers in the boxes.

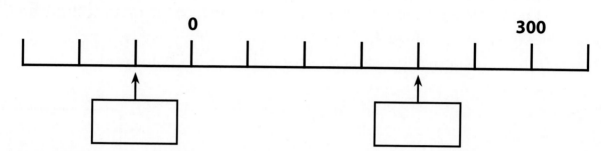

0

300

12 Circle **two** numbers that have a difference of **3**.

5   −2.5   0   −2   0.5

13 Write down all the valid Roman numerals you can find using one or more of the numerals **I**, **V** and **X** only once each.

....

14 Find a number between **1** and **2000** which has more than 10 letters when written using Roman numerals.

15 Use the digits **1, 1, 2, 3, 6**.
Find the number greater than **50,000** which has:

The same **tens** and **thousands** digits

    AND

the **tens of thousands** digit is **double** the **ones** digit.

16 A large department store has **6** floors.
In the basement, there are **3** levels of car park.

On which level are:

Level	Department
5	**Electrical Goods**
4	**Men's Clothes**
3	**Children's Clothes**
2	**Shoes**
1	**Women's Clothes**
0	**Beauty**
−1	**Car Park 1**
−2	**Car Park 2**
−3	**Car Park 3**

a Men's Clothes?

b Electrical Goods?

c Shoes?

Mr Hudson parks in Car Park **2**.
He goes up to Women's Clothes but then goes back to his car.
He then goes back up to Men's Clothes.

d How many floors has he visited in total?

How many floors apart are:

e **Men's Clothes** and **Beauty**?

f **Car Park 3** and **Car Park 1**?

g **Electrical Goods** and **Car Park 3**?

17 Find the **first** number greater than **34,695** where:

a its **thousands** and **ones** digits are the same.

b its **thousands** and **tens** digits are the same.

c it is divisible by 9.

18 **a** Use the digits **0, 0, 3, 5, 6** to make 4 numbers between **50,000** and **60,000**.

b Write your numbers in descending order.

c Find the difference between the largest and smallest numbers.

19 Mark the numbers **−0.6, 1.2 and −1.2** on this number line.

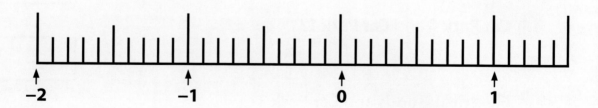

−2　　　　−1　　　　0　　　　1

20 What is the difference between the temperatures on these thermometers?

 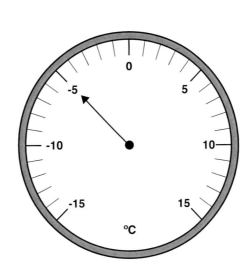

°C

21 Complete the table below.

	rounded to the nearest 10	rounded to the nearest 100	rounded to the nearest 1000	rounded to the nearest 10,000
46,513				
23,042				
84,639				
181,315				
65,065				

22 Tom goes to the fair. He pays **£1** to have **three** goes at the Hoopla. He has **three** rings for each go.

a Write down each Hoopla total.

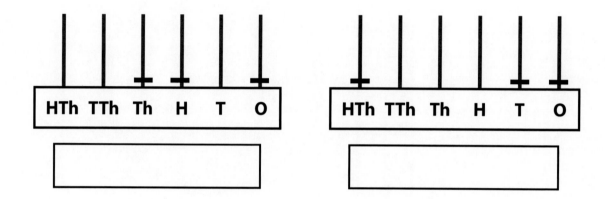

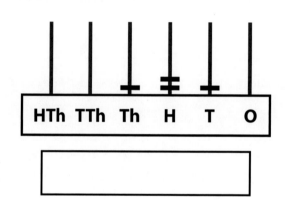

b What is the **biggest** number that can be made using three hoops?

c What is the **smallest** number that can be made using three hoops?

23 **a** Round each of these numbers to the nearest **1000**.

12,345	67,765	99,565

b Find the difference between each original number and the rounded number.

---

24 **Four** girls took part in a **100 m** race.
The girl in Lane 1 finished first.
A photograph was taken showing where each of the other girls was as the girl in Lane 1 finished.
Estimate the distance the girls had each run when the photograph was taken.

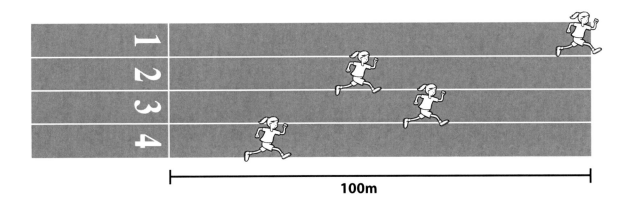

100m

Lane 2 = [ ] m   Lane 3 = [ ] m

Lane 4 = [ ] m

Name ....

NUMBER

Addition and subtraction

These are all about addition and subtraction!

Please note that the Mastering sections for **Number - Addition and subtraction** and **Number - Multiplication and division** are combined following the Multliplication and division section in the book.
Children will need to apply their reasoning skills to link these areas.

2.10.23

Add whole numbers with more than 4 digits using a formal written method where appropriate

1 The cook makes **1085** chicken lunches and **812** vegetarian lunches in one month. How many lunches has he made altogether?

1897 ✓

2 The builder uses **2348** bricks to build Wall A and **1732** to build Wall B. How many bricks did she use in total?

4080 ✓

3 The mischievous monkeys love bananas. On Monday, they eat **1238**. The monkeys eat **2347** bananas on Tuesday. How many bananas have they eaten in total?

3585 ✓

4 Tom is trying to work out the answer to **3583** add **2438**. What should his answer be?

6021 ✓

5 The blue lorry is carrying **3482** boxes of cereal. The green lorry holds **5879** boxes. How many boxes of cereal are there altogether?

9361 ✓

6 What is the total of **16,743** and **7762**?

24,505 ✓

7 On a hot beach, there were **12,489** people sunbathing on Monday. On Tuesday, there were **14,723** people on the beach. How many people were there altogether on Monday and Tuesday?

27212 ✓

8 Add together **32,485** slimy snails and **48,643** wiggly worms. How many snails and worms are there altogether?

81128 ✓ checked

Name ....

Add whole numbers with more than 4 digits using a formal written method where appropriate

1 At a cricket match, there are **1265** people supporting the Shorthrow team and **1726** supporting the Throwlong team. How many people are watching the game altogether?

2991

2 **2168** women and **2453** men live in Coughlee Village. How many adults live in the village in total?

4621

3 One huge tropical fish tank has **1226** fish and another has **1189**. How many fish are there in total?

2415

4 The pie man sells **1535** meat pies and **1695** potato pies at the rugby match. How many pies has he sold altogether?

3230

5 Mr Davies asks Class 5 to work out **4379** add **3872**. What should their answer be?

8251

6 Bilal works out the answer to **5349** add **7899**. What answer should he write down?

13248 *checked*

7 There is a big sale at the Hafford Shopping Centre. On the first day, **12,478** people went shopping. On the second day, **18,653** people visited. How many people went shopping in total?

31131

8 What is the total of **245,753** and **347,789**?

593542

Name ....

Add whole numbers with more than 4 digits using a formal written method where appropriate

1 The baker made **1342** jam tarts and **1446** lemon tarts. How many tarts did he bake altogether?

2 In the forest, there are **1832** oak trees and **2167** sycamore trees. What is the total number of oak and sycamore trees in the forest?

3 The postman delivers **2348** letters on Monday. On Tuesday, he delivers **4633**. How many letters does he deliver altogether?

4 Charlie scores **3523** points on the Gorilla Crush computer game. Sadia scores **6477** points. How many points do they score altogether?

5 In one week, a supermarket sells **2679** apples and **2543** bananas. How many apples and bananas are sold altogether?

6 Ayesha is trying to work out the answer to **4873** add **8547**. What should her answer be?

7 At the football match, there are **28,186** Hottingham fans and **21,364** Pelsea fans. How many fans are watching the football game altogether?

8 In the lake, there are **136,748** trout, **15,463** pike and **42** salmon. How many fish are there altogether?

Name ....

Subtract whole numbers with more than 4 digits using a formal written method where appropriate

1 Tom has collected **1533** animal stickers. Lola has collected **412** less. How many animal stickers does Lola have?

1121

2 At the Crazy Circus, there are **2845** people watching. **2546** watching are children. How many adults are watching the circus?

299

3 There are **3532** tadpoles in the pond. By Tuesday, **2348** have turned into frogs. How many tadpoles are left in the pond?

1184

4 Ayesha has forgotten how to carry out formal written subtraction. How should she work out **3452** subtract **2568** using a formal written method?

884

5 Charlie collects comics. He has **2126** comics. How many more does he need to collect to reach **3450**?

1324

6 The Brainy Biffs have **3547** points and the Clever Clogs have **2489**. What is the difference between the points?

1058

7 Zany Zoo buys **150,500** bananas a month. The baboons eat **97,645**. How many bananas are left for the other animals?

52855

8 I think of a number then add **659,879**. The answer is **732,662**. What was my number?

72783

25.10.23

Subtract whole numbers with more than 4 digits using a formal written method where appropriate

1 Charlie scored **2572** points on the Dancing Feet game. Sadia scored **1351** points. How many more points did Charlie score than Sadia?

2221

2 There are **3257** people at the hockey match. This is **1138** people more than last week. How many people were at the match last week?

2119

3 There were **5834** guests staying at the holiday resort during one week. The next week, there were **3425** guests staying. How many more people stayed at the resort during the first week?

2409

4 Bilal is struggling to work out the answer to **4532** subtract **2643**. What should his answer be?

1889

5 **23,673** people enter a talent show. **11,489** are knocked out in the first round by the judge Simon Cowbell. How many people are left in the talent show?

12184

6 A bumper box of sequins holds **23,453** silver sequins and **21,567** gold sequins. How many more silver sequins are in the box?

1886

7 I think of a number then add **138,659**. The answer is **312,234**. What was my number?

173575

checked

8 What is the answer to **600,000** subtract **145,893**?

454107

W27   10-3-24

NUMBER - Addition and subtraction

Year 5

Subtract whole numbers with more than 4 digits using a formal written method where appropriate

1 In the shoe warehouse, there are **1848** pairs of shoes. **737** pairs are trainers. How many are not trainers?

1111 pairs

2 Show how Ayesha should use a written method to find the answer to **2782** subtract **1456**.

```
  2782
- 1456
------
  1326
```

3 The Wippy man has **2742** ice creams in his van. He sells **1762** on a really hot weekend. How many does he have left to sell?

```
  2742
- 1762
------
  0980
```

4 Work out the answer to **6483** subtract **2294**.

4189

5 Sadia is trying to work out the answer to **5643** subtract **2764**. What should her answer be?

```
  5643
- 2764
------
  2879
```

2879

6 **18,755** people go to watch the pop band One Diversion. **7889** people leave the event early because of a power cut. How many people are left?

10866

checked

7 What is the answer to **142,543** subtract **38,667**?

103876

8 Explain to Tom how to do column subtraction using numbers or words or both.

```
  142543
-  38667
-------
  103876
```

© Copyright HeadStart Primary Ltd   **40**   Name ....

Add and subtract whole numbers with more than 4 digits using a formal written method where appropriate

1 A bakery sold **3347** loaves on one day and **4532** the next day. How many loaves were sold altogether?

2 Charlie had to find the difference between **6574** and **3459**. What should his answer have been?

3 Tom has forgotten how to use formal written addition. Show him how to add **3672** add **2328**.

4 Lola's teacher asks her to find the difference between **63,467** and **32,279**. What should her answer be?

5 What is the difference between **67,123** and **45,679**?

6 There are **34,583** Reverton fans and **25,542** Sarsenel fans at the football match. How many fans are there altogether?

7 What do you need to add to **241,647** to get an answer of **321,336**?

8 What is the sum of **2567**, **10,042** and **324,356**?

Name ....

Add and subtract whole numbers with more than 4 digits using a formal written method where appropriate

1 Bilal scored **3453** on a computer game. Lola scored **5234**. How many did Bilal and Lola score altogether?

2 In a counting competition, Ayesha counted all the way up to **4256**. Tom could only manage to count up to **3128**. How many more did Ayesha count than Tom?

3 **34,256** people managed to get a ticket for the U3 concert. **13,178** people did not get a ticket. How many would have been watching if they had all got tickets?

4 Mrs Johnson won **£15,265** on the lottery. Mr Brown won **£12,876**. How much more did Mrs Johnson win than Mr Brown?

5 Find the difference between **452,432** and **235,678**.

6 Ayesha thinks of a number and subtracts **656,346**. Her answer is **344,789**. What was her first number?

7 Charlie wanted to check that **244,647** add **484,456** was equal to **729,103**. Show a subtraction calculation he could use to check.

8 In a year, a bakery makes **234,321** chocolate cakes, **134,873** vanilla cakes and **23,476** apple pies. How many items do they bake altogether?

42

Name ....

Add and subtract whole numbers with more than 4 digits using a formal written method where appropriate

1 Miss Bradshaw bought **2247** fizzy gem sweets in a big tub. She ate **1134** in 3 months. How many did she have left?

2 There are **2342** marbles in a box. Year 5 take **1223** marbles out. How many marbles are left?

3 Tom wanted to use a written column method to calculate **3452** add **4358**. Show the method that he could have used.

4 There are **15,345** books in the library. **3467** books are taken out. How many books are left in the library?

5 Ten primary schools buy a total of **12,562** pencils and **23,548** coloured pencils for the new school year. How many pencils have they bought altogether?

6 What is the difference between **234,675** and **145,794**?

7 In the rainforest there are **136,743** hummingbirds, **254,367** finches and **124** harpy eagles. How many birds are there in total?

8 I think of a number then add **512,234**. The answer is **7,245,000**. What was my number?

Add numbers mentally with increasingly large numbers

1 Miss Smith bakes **50** chocolate cakes and **100** iced cakes for the cake sale. How many cakes has she baked altogether?

2 There are **161** pupils at Harper Grove Primary and **40** teachers and staff. How many people are there in total?

3 In the garden, Grandma has grown **129** red roses and **221** white roses. How many roses has she grown altogether?

4 At the cinema, there are **450** people watching a comedy film and **465** people watching a superhero film. How many people are in the cinema?

5 Ayesha adds **520** and **345**. What should her answer be?

6 Tom works out the answer to **2320** plus **3459**. His answer is **5780**. Is he correct? Explain your answer.

7 There are **2548** red balls and **1452** blue balls in the ball pond. How many balls are there in total?

8 Charlie thinks of a number. He subtracts **345**. His answer is **427**. What number did he think of?

Subtract numbers mentally with increasingly large numbers

1 Mrs Harrison made **90** sandwiches for the school trip to the beach. The seagulls swoop down and steal **40** sandwiches. How many sandwiches are left?

2 Bilal has **200** smiley stickers. Lola has **59** less. How many stickers does Lola have?

3 There are **131** eggs in the incubator. **95** eggs hatch into chickens. How many eggs are left in the incubator?

4 Charlie collects comics. He has **268** comics. How many more does he need to collect to reach **370**?

5 Two teams are having a quiz. The Wizz Kids have **647** points and The Brain Busters have a score of **325**. What is the difference between their scores?

6 Kim, the kangaroo, jumps **2450** times a day. Joey, the kangaroo, jumps **1200** times less. How many times a day does Joey jump?

7 The café bought **3280** tea bags. They used **1279**. How many tea bags were left?

8 What is the difference between **5004** and **3499**?

Add and subtract numbers mentally with increasingly large numbers

1 On a book case, there are **110** books on one shelf and **90** books on another. How many books are there altogether?

2 Lola collected **155** loom bands. She gave **45** of them to her sister. How many bands did she have left?

3 Ayesha uses a mental method to work out **399** subtract **120**. What should her answer be?

4 Mr Right writes this calculation on the board: **467** add **532**. What should the answer be?

5 What is the difference between **768** and **345**?

6 An ASDA lorry is carrying **3480** tins of beans. A Tesco lorry is carrying **5620** tins. How many tins of beans are there altogether in both lorries?

7 What is the total of **6742** and **175**?

8 Grandad has forgotten how to work out calculations mentally. Explain how he could work out **2469** subtract **1201** using a mental method.

Name ....

Add and subtract numbers mentally with increasingly large numbers

1 At the school fair, there are **140** children and **60** adults. How many people are at the fair in total?

2 There are **135** cakes in the bakery. **75** are sold at lunchtime. How many cakes are left?

3 Charlie thinks of a number then adds **159**. The answer is **300**. What was his number?

4 There are **227** blue soft balls in a play area and **253** red. How many soft balls are there in the play area?

5 Mrs Stowe asks her class to work out **880** subtract **352**. What answer should they say?

6 In the art cupboards, there are **1251** colouring pencils. Mrs Rajan takes **249**. How many pencils are left?

7 There are **4578** cherries on the trees. The birds eat **2399**. How many cherries are left?

8 Work out the answer to **3477** add **2352**.

Name ....

Use rounding to check answers to calculations

1 **81** girls and **52** boys are watching the school play. Estimate the number of children that attended, by rounding each number to the nearest **10**.

130

2 Which calculation is a good estimate for **299** add **297**?

400   (**600**)   **500**

3 Tom worked out that the answer to **347** subtract **131** was **278**. Is he correct? How could he use rounding to check his answer? *he is*

wrong as he has added the 47 and 31 and subtracted 300 and 100

4 A train has two carriages, one holding **159** passengers and the other holding **162**. Ayesha says there are **301** passengers altogether. Is she correct? Explain your answer, using rounding. *No as 162 rounded as 160 and 159 rounded as 160, so if you add them you would get 320*

5 How could you estimate the answer to **1898** add **2243**?

4100   *1900 +2200 4100*

6 Charlie writes down the calculation **2298** subtract **1221** on the blackboard. How could he estimate his answer? *by rounding 2298 to 2300 and rounding 1221 to 1200 and adding them = 1100*

7 Bilal says that **2456** plus **1550** to the nearest **100** is **4000**. Is he correct? Explain your answer. *yes as 2456+1550 = 4006 and 4006 rounded to the nearest 100 is 4000*

Name ....

Use rounding to check answers to calculations

1 A train has two carriages, one holding **39** passengers and the other **53**. Tom says that there are **101** people altogether. Use rounding to explain how you know he is wrong.

2 Sadia has **168** stickers and Bilal has **163**. Estimate the total number of stickers that they have altogether, by rounding each number to the nearest **10**.

3 Mr Bolt has **£248**. He spends **£82** on some trainers. To the nearest **£10**, how much money does he have left?

4 There are **176** children and **45** members of staff at Biff Street Primary. The headteacher wants to buy each person a pencil. She orders **220** pencils. Has she ordered enough? Use rounding to explain your answer.

5 Bilal works out the answer below.

$$1347 + 2412 = 3859$$

Is he correct? Use rounding to check his answer.

6 A farmer grows **4679** cabbages. The slugs ruin **1483**. How many cabbages are left? Work out the answer and use rounding to check.

Name ....

When rounding, determine levels of accuracy in the context of a problem

1 There are **29** children in Class 5. Mr Rosthorn, the class teacher, rounds **29** to the nearest **ten** and buys **30** lollies as a reward for his class. Did he buy enough lollies? Explain your answer.

2 The cat eats **35** meals a week. Her owner rounds this and buys **30** meals. Will the cat have enough food to eat for the week? Explain your answer.

3 Dad worked out it would take about **65** rolls of wallpaper to decorate every room in the house. He decided to round **65** to the nearest **100** and so he bought **100** rolls of wallpaper. Was this a sensible thing to do? Explain your answer.

4 There are **159** children who have school lunches. Mrs Burnitt, the school cook, rounded this and made **200** lunches. Was this a waste of food? Explain your answer.

5 There are **408** children at Redtree School. The secretary needs to print a letter for each child. She rounds **408** to **400** and prints **400** letters. Was this a sensible use of rounding? Explain your answer.

Solve addition and subtraction multi-step problems in contexts, deciding which operations to use and why (money)

1 Lola had **£25**. She bought a new DVD for **£12** and a hair band for **£4**. How much money did she have left?

2 Tom earns **£37** on Saturday. On Sunday he earns **£25**. How much money did he earn altogether?

3 For Ayesha's birthday her grandma gives her **£25**. Her grandad gives her **£20**. How much money does she have left if she buys some shoes for **£15**?

4 Charlie goes shopping and spends **£55** on some trainers, **£28** on a t-shirt and **£36** on some jeans. How much money has he spent altogether?

5 Bilal buys a CD for **£6** and a hand-held computer game for **£10.50**. How much change does he receive from **£20**?

6 What is the total cost of a yo-yo costing **£2.75**, a DVD costing **£14** and a poster costing **£2.75**?

7 Sadia has **£6.50**. She buys an ice cream for **£1.50** and a comic for **£1.75**. Does she have enough to buy a game for **£3.50**? Explain your answer.

Name ....

Solve addition and subtraction multi-step problems in contexts, deciding which operations to use and why

1 There are **250** children eating their lunch. **45** leave to go to the playground. **12** more children arrive to eat their lunch. How many children are eating now?

2 The Year 5 class has **33** pupils. Year 6 has **4** more pupils. How many pupils are in Year 5 and 6 altogether?

3 Lola's book has **350** pages. She reads **60** pages on Tuesday and **36** on Wednesday. How many pages does she have left to read?

4 **Two hundred and forty six** people go to the country park on Sunday. This is **54** fewer people than went on Saturday. How many people went to the country park on Saturday and Sunday altogether?

5 3 children need **144** gold stars between them to win a prize. Sadia and Ayesha have **24** gold stars each. Bilal has **27** gold stars. How many more gold stars do they need to win a prize?

6 Charlie has **£100**. He spends **£18** on a game and **£45** on a new coat. He then spends **£22** on a t-shirt. How much money has he spent in total? Does he have enough money to buy a toy car for **£12**? Explain your answer.

7 Bilal thinks of a number. He adds **437** and then subtracts **67**. His answer is **852**. What was his first number?

Name ....

NUMBER

Multiplication and division

These are all about multiplication and division!

NUMBER - Multiplication and division

Identify multiples

1 Mrs Yan's class thought the numbers **12** and **24** were both multiples of **6**. Are they correct? Explain your answer. *& yes* $6 \times 2 = 12$

$6 \times 4 = 24$

2 Which of the numbers below are multiples of **2**?

(12)   (46)   33   21   (70)

3 Tom wanted to find **3** multiples of **4**. What could his answer be?

4, 8, 12

4 Bilal says, "The number **45** is a multiple of **5**." Is he correct? Explain your answer. *yes because is a number ends in 0 or 5 its in the 5 X table*

5 Is it true that all multiples of **3** are also multiples of **6**? Explain your answer using examples. *True 3 x 4 = 12 false because 3 isn't a multiple of 6*

6 Charlie says, "The number **243** is a multiple of **9**." Is he correct? Explain your answer. *yes because the digit sum is 9*

7 Find **2** numbers for which **15** is a multiple.

3 and 5

8 Lola needed to find out whether the numbers below would divide equally by **3**. Which ones did she find to be divisible by **3**?

(345)   344   (546)   223   *checked*

53

Name ....

18.10.23

Identify multiples

1 Which number is a multiple of **two**: **34** or **45**?

34

2 Which of the numbers below are multiples of **5**?

(35)   21   (40)   (55)   (600)   4

3 Sadia says all multiples of **6** are even. Is she always correct? Give some examples to support your answer. *true*

6,12,18,24,30,36,42,48,54,60,66,72,78

4 Tom thought that all the multiples of **2** were also multiples of **4**. Is he correct? Explain your answer using examples. *yes/no as*
2 isnt in the 4x table

5 Miss Coffee asks her class to find all the multiples of **7**, smaller than **50**. Which numbers should they write down?

7,14,21,28,35,42,49

6 Ayesha says, "The sum of **two** even numbers is always a multiple of **4**." Is she correct? Give some examples to support your answer. *No as*
2+4 =6 and 6 is not in the 4 x table

7 Sadia is trying to work out all the multiples of **9**, bigger than **30** but smaller than **60**. Which multiples should she write down?

36, 45, 54,

8 Tom wants to know if all multiples of **9** are also multiples of **3**. Give some examples to show whether this is true or not. *yes as 9x2=18*
3x6=18

NUMBER - Multiplication and division

Identify factors, including factor pairs of a number and common factors of 2 numbers

1 Identify all the factors of the number **12**.

1, 2, 3, 4, 6, 12

2 Ayesha has worked out that **2** is a factor of the number **6**. What other number would make a factor pair?

2 and 3   1 and 6

3 Lola says, "Every number has itself (the number) and **one** as a factor." Is she correct? Explain your answer. yes as one times any number is to that number .e.g. 1×17=17

4 Tom uses his multiplication knowledge to help find the factors of **22**. What numbers should he write down?

1, 2, 11, 22

5 Ayesha writes down the number **42**. She has worked out that **7** is a factor. What other number would make a factor pair?

checked

7 and 6   1 and 42

6 Which of the numbers below are common factors of **36** and **24**?

(4)   5   (6)   (2)   (1)   9

7 Work out all the factors of **85**.

1, 5, 17, 85

8 Work out all the factor pairs of **96**.

1×96, 2×48, 3×32, 4×24, 6×16, 8×12

1, 2, 3, 4, 6, 8, 12, 16, 24, 32, 48, 96

Name ....

NUMBER - Multiplication and division

Identify factors, including factor pairs of a number and common factors of 2 numbers

1 What are all the factors of the number **30**?

1, 2, 3, 5, 6, 10, 15, 30

2 Bilal says, "**5** is a factor of **12**." Is he correct? Explain your answer. No as

12 5 isn't in the 5 times table

3 Look at the pairs of numbers below. Circle the factor pair of **18**.

(6 and 3)   5 and 2   1 and 12

4 Sadia is looking for the factor pairs of **45**. She knows that **5** is a factor. What is the other number in the factor pair?

5 and 9   ~~1 and 45~~   ~~3 and 15~~

5 Which of the numbers below are common factors of **45** and **30**?

2   6   (5)   9   (3)   checked

6 Ayesha works out that **4** is a common factor of **44** and **60**. Write down **two** more numbers which have a factor of **4**.

16   32

7 Mr Smith asks his class to write down all the factor pairs of **72**. What numbers should they write down? 1x72, 2x36, 3x24, 4x18, 6x12, 8x9

1, 2, 3, 4, 6, 8, 9, 12, 18, 24, 36, 72

8 Write down all the factor pairs of **140**. 1x140, 2x70, 4x35, 5x28, 7x20, 10x14

1, 2, 4, 5, 7, 10, 14, 20, 28, 35, 70, 140

Know and use the vocabulary of prime numbers, composite (non-prime) numbers and identify prime and composite numbers

1 Tom thinks that **7** is a prime number. Is he correct? Explain your answer.

2 Charlie believes that composite numbers have more than **2** factors. Is he correct? Explain your answer.

3 Ayesha is trying to write down all the prime numbers from **1** to **10**. What should she write down?

4 Lola says, "A number that can be divided evenly is a composite number." Is she correct? Explain your answer.

5 Which of the numbers below are composite numbers?

41   **63**   **56**   **68**   **131**

6 What factors do prime numbers always have? Explain your answer using an example.

7 Which are the missing prime numbers in this sequence?

101   **103**   [   ]   **109**   [   ]

8 Tom's dad has forgotten what a prime number is. Can you explain to him? Use examples to support your explanation.

Know and use the vocabulary of prime factors

1 Tom writes down the numbers **12** and **13**. Which number is a prime number?

2 Mrs Thomas asks her class to explain what a prime factor is. How should they do this?

3 What are the prime factors of **24**?

4 Bilal thinks that the prime factors of **18** are **3** and **6**. Is he correct? Explain your answer.

5 What are the prime factors of **39**?

6 Mrs Shaw asks her class to explain prime factorisation. How should they do this?

7 Complete the prime factorisation process for the number **20**.

8 Lola is struggling to work out how to complete prime factorisation for the number **84**. What should she write down?

Name ....

(handwritten: 4-10-23)

Multiply numbers up to 4 digits by a one-digit number, using a formal written method

1 There are **14** apples in a box. Tom's mum buys **three** boxes. How many apples has she bought? Show your working out.

(handwritten working: 14 × 3 = 42)

42

2 Bilal wanted to use a written method to calculate **23 x 6**. Show the method that he could have used.

(handwritten working: 23 × 6 = 138)

138

3 There are **16** pens in a box. An office buys **8** boxes. How many pens do they buy?

(handwritten working: 16 × 8 = 128)

128

4 A taxi driver takes **4** people in every taxi ride. On Saturday and Sunday, he makes **88** journeys. How many people did the taxi driver take altogether?

352

5 Each aeroplane carries **157** passengers. There are **8** planes which set off from Hatwick in the morning. How many passengers are there in total?

1256

6 The school cook wanted to give each child **4** sandwiches for their school trip. There are **974** children in the high school. How many sandwiches would she need?

3896

7 There are **7** sweet stalls at the market. If each stall sold **1235** sweets, how many sweets would be sold altogether?

(handwritten: checked)

8645

11-10-23

Multiply numbers up to 4 digits by a two-digit number, using a formal written method

1 There are **22** children in Year 5. Each child is given **12** house points. How many house points are given in total?

264

2 Carrots come in packs of **15**. The chef buys **18** packs for his restaurant. How many carrots has he purchased?

270

3 Miss Bradshaw bought **twelve** packets of chocolate biscuits. Each packet contained **34** biscuits. How many biscuits did she buy altogether?

408

4 Lola got **£30** each week from working at the café on Saturdays. She saved this money for a year. (A year is **52** weeks). How much money did she save in total?

£1560

5 A stand in a sports stadium has **245** rows of **87** seats. How many seats are there in the stand?

21315

6 Each bag holds **475** raisins. A shop buys **90** bags. How many raisins are there altogether?

42750

7 Mr Malik is saving up for a new car. He saves **£1345** each month. He saves up for **12** months. How much money does he save altogether?

16140   checked.

Multiply and divide numbers mentally drawing upon known facts

1 Bilal knows that **21** divided by **7** is **3**. He uses this knowledge to work out that **210** divided by **70** is **3**. Explain how he knew this.

2 There are **11** players in a football team. There are **15** teams in the tournament. How many players are there altogether? Show how you can work this out mentally.

3 Charlie knows that **105 ÷ 7 = 15**. He uses this to work out **15 × 7**. What would his answer be?

4 Use the facts from Question **3** to calculate **150 × 7**.

5 Mrs Cole says **63** multiplied by **7** equals **441**. Use this information to work out **441** divided by **7**.

6 Bilal knows that **6 × 9 = 54**. Explain how he could use this calculation to work out **60 × 90**.

7 Ayesha wants to work out the answer to **9 × 17**. She knows her **9** times table up to **12 × 9**. Explain how she can use this knowledge to find the answer.

8 Explain a different way Ayesha could work out **9 × 17**.

Name ....

Multiply and divide numbers mentally drawing upon known facts

1 Sadia knows that **4 × 3 = 12**. She uses this work to out **4 × 30**. What should her answer be?

2 Use information from Question 1 to answer **400** multiplied by **3**.

3 If **56** divided by **7** equals **8**, what is the answer to **560** divided by **7**? Explain how you know.

4 Bilal knows that **18 × 9 = 162**. He uses this fact to work out **162 ÷ 18**. What should his answer be?

5 If **5 × 11** equals **55**, what is the answer to **110 × 5**?

6 Which multiplication statement below would help you work out the answer to **1089 ÷ 99 = 11**?

999 × 11   **99 × 10**   **11 × 99**

7 Sadia knows that **8 × 11 = 88**. She uses this to work out **880** divided by **11**. What should her answer be?

8 **108 ÷ 9 = 12**. Explain how you can use this division fact to work out the answer to **90 × 12**.

62   Name ....

Divide numbers up to 4 digits by a one-digit number, using the formal written method of short division

1 Bilal has **57** t-shirts. His mum asks him to put them equally into his **3** drawers. How many t-shirts does he put in each drawer?

2 There are **72** grapes in a bag. How many people can have **8** grapes each?

3 A big box of biscuits has **275** biscuits in **5** layers. How many biscuits are in each layer?

4 Tom wanted to organise **336** marbles into bags with **7** marbles in each. How many bags of **7** could he make?

5 Sadia thinks of a number and multiplies it by **4**. Her answer is **676**. What was her number?

6 Mr McKay, the headteacher, wanted to organise all the children at his high school into **3** equal teams for sports day. There were **995** children in the school. How many would there be in each team? How many children would be left over?

7 Mum puts **£776** into a bank account and Dad adds another **£2049**. They share the money equally between their **five** children. How much does each child get?

Name ....

Divide numbers up to 4 digits by a one-digit number and interpret the remainder appropriately for the context

1 Lola has **28** stickers. She shares them equally between her **5** friends. How many stickers will each of her friends get? How many stickers will Lola have left?

2 Mrs Bailey has **43** bean bags. She shares them out between **8** children. How many bean bags does each child have? How many are left over?

3 There are **43** biscuits in a box in the staffroom. **Seven** teachers have **5** biscuits each but Mr Jones and Mrs James have only **4** each. Explain how this happened.

4 There are **29** children in the hall. Mr Bean asks the children to get into **four** equal teams. Is there a problem with this? Explain your answer.

5 There are **188** children going on a school trip. They are going in groups of **6**, with **one** adult per group. One group will have less than **6** but still needs **one** adult. How many people are going on the trip altogether?

6 A bag needs **6** sweets to be full. There are **1345** sweets. How many full bags can be made? How many sweets are left over?

7 **2026** people attend the theatre. There are **9** seats in a row. How many full rows are there? How many more people are needed to fill another row?

Name ....

W 46
20.7.24

checked

Multiply and divide whole numbers and those involving decimals by 10, 100 and 1000

1 There are **300** bouncy balls. Mr Hughes shares the balls equally between **10** children. How many bouncy balls does each child get?

30

2 At a fair, **100** children went on the ride. Each ride cost **£2**. How much was spent on the ride altogether?

£200

3 Lola has **1000** trading cards. She gives **one tenth** of the cards to Ayesha. How many cards does Ayesha get?

100

4 Bars of chocolate cost **32p**. They are sold in packs of **10**. How much does a pack of **10** cost?

£3.20

5 At the Grand Garden Centre, there are **1000** flowers. Each flower has **6** petals. How many petals are there altogether?

6000

6 There are **92,000** paper clips ready to be packed into boxes. Each box contains **1000** paper clips. How many boxes could you fill, if you used all the paper clips?

92

7 The shopkeeper paid **£360** for a delivery of soap. **One hundred** boxes were delivered altogether. How much did the shopkeeper pay for **one** box of soap?

£3.60

8 A school charges each family **£7.50** for the children's trips. There are **100** families. How much does the school charge altogether?

£750

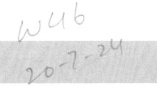

W46
20-7-24

Multiply and divide whole numbers and those involving decimals by 10, 100 and 1000

1 Bilal is trying to work out the answer to **534** multiplied by **10**. What should his answer be?

5340

2 A school has **1000** pupils. Each pupil needs **12** coloured pencils. How many pencils does the school need to order?

12000

3 The school cook makes **1200** fruit snacks for the school. There are **10** classes. How many fruit snacks can each class have?

120

4 **Thirty thousand** people went to a football match. **One tenth** of them were children. How many children were at the match?

3000

5 Lola has saved **£25** for her holiday to Portugal. She divides the money equally between the **10** days she is on holiday. How much money does she have each day?

£2.50

6 Sadia wanted to work out **4357** divided by **100**. What should her answer be?

43.57

Checked

7 Mr Bulat pays **£115** every week for **100** weeks to pay off a loan for his car. How much did his car cost altogether?

£11500

66

Name ....

W46 26.7.24 Checked

Recognise and use square numbers

1 Which of the following is a square number?

5   (4)   6

2 Class 5 are learning about square numbers. Which number have they missed out in this sequence?

9   16   25   36

3 Bilal says, "**49** is a square number." Is he correct? Explain your answer.

Yes as $7^2 = 7 \times 7 = 49$

4 What is **10** squared?

100

5 What is the square root of **64**?

8 ✓

6 Lola has forgotten what a square number is. Can you explain using words or examples?

a number timesed by itself

7 Circle all the square numbers below.

(49)   (9)   35   (81)   62   (121)

8 Tom thinks that the square of **12** is **134**. Is he correct? Explain your answer.

no as $12^2 = 12 \times 12 = 144$

W46
20-7-24

Recognise and use cube numbers

1 Which of the numbers below is a cube number?

8　　**9**　　**10**

(8 is circled)

Checked ✓

2 Class 5 are learning about cube numbers. Which number have they missed out in this sequence?

27　　　**64**　　　| 125 |

3 Ayesha says, "**216** is the cube number of **6**." Is she correct? Explain your answer.

yes as $6^3 = 6 \times 6 \times 6 = 216$

$36 \times 6 = 216$

4 Write down the cubed number of **5**.

$5^3 = 125$

5 Lola is trying to work out the answer to 8^3. What should her answer be?

512

6 Charlie says, "The cube root of **25** is **5**." Is he correct? Explain your answer.

No as $5 \times 5 \times 5 = 125$

7 Sadia says **1000** is a cube number. Is she correct? Explain your answer.

yes as $10 \times 10 \times 10 = 1000$

8 What is the cube root of **125**?

5

W 46
23-7-24

Checked

Solve problems involving multiplication and division, including using knowledge of factors and multiples, squares and cubes

1 Charlie said he had **5²** trading cards. How many trading cards did he have?

25

2 What number, between **33** and **38**, is a multiple of **4**?

36

3 What number, between **20** and **30**, is a cube number?

27

4 In Mr Pott's garden, there are **7** rows of tomato plants. Each row has **435** tomato plants. How many tomato plants are in the garden altogether?

3045

435
7
3045

5 There are **386** apples that need to be put into boxes containing **5** apples per box. How many full boxes of apples can be made altogether? How many apples are left?

77 r1

6 Tom has forgotten what a prime number is. Can you explain this to him, using an example?

a number only divisibly by 1 and itself

7 Charlie cuts **10** pieces of string from a ball. Each piece is **1.7 metres** long. What would be the total length of the string if the pieces were laid end to end?

17m

8 Use your knowledge of **7 × 9** to work out **700 × 90**.

63000

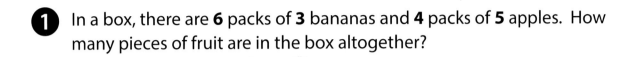

W46
23-7-24
checked

Solve problems involving addition, subtraction, multiplication and division and a combination of these

1 In a box, there are **6** packs of **3** bananas and **4** packs of **5** apples. How many pieces of fruit are in the box altogether?

38

2 A bag of sweets contains **10** sweets. Charlie has **2** bags of sweets. Bilal has **3** bags of sweets. How many sweets do they have in total?

50

3 Mr Otu cycles **9** miles a day for **12** days. His friend, Mr Samsung, cycles **15** miles for **4** days. Who has cycled further and by how many miles?

Mr Otu Rby 48miles

4 Sadia walks **2** miles every day. Ayesha walks **3** miles every day. How many more miles does Ayesha walk in a week?

7miles

5 All the children in the school are going litter picking. They are working in groups of **6**. If there are **142** children in the school, how many litter picking groups of **6** will there be? How many children will be left over?

23 r 4

6 Lola swam **20** lengths a day for **13** days. Charlie swam **16** lengths for **17** days. Who swam further? By how much?

Charlie by 12

7 Sadia thinks of a number and multiplies it by **6**. She then subtracts **12**. Her answer is **138**. What was the number she thought of first?

25

Name ....

Solve problems involving multiplication and division, including scaling simple fractions and problems involving simple rates

1 The sycamore tree in Mrs Green's garden is **18 metres** tall. Her willow tree is **half** as big as this. How tall is the willow tree?

9 m

2 Ayesha built a **40 cm** tower out of building blocks. Lola built a tower **a quarter** as big as Ayesha's tower. How tall was Lola's tower?

10 cm

3 The length of the school field is **50 metres**. Year 5 plant trees over $\frac{1}{5}$ of the length. Over how many **metres** are trees planted?

10 m

4 Sadia is saving to go on holiday. The holiday costs **£200**. Over a year, she saves **one and a half** times the cost of the holiday. How much does she save?

£300 ✓

5 Mrs Wallace borrowed **£400** from the bank to buy a new laptop. She paid back **one and a quarter** times as much to the bank. How much did she pay back?

£500

6 Mr Smith has borrowed **£5000** from the bank to buy a new car. He had to pay back **£5000** plus $\frac{1}{10}$ of **£5000**. (This is called interest.) How much did he have to pay back altogether?

5500

Name ....

MASTERING

Addition, subtraction, multiplication and division

Use your best reasoning skills to solve these!

MASTERING - Addition, subtraction, multiplication and division   **Year 5**

checked

5489
4247
9736

1 Use the inverse operation to check these calculations.
Tick the ones that are correct.

5687 + 2458 = 8145 ✓

9736 − 5489 = 4247 ✓

a $8145 - 2458 = 5687$

b $4247 + 5489 = 9736$

2578 × 8 = 20624 ✓

5274 ÷ 9 = 586 ✓

c $20624 \div 8 = 2\cancel{8475}1579$

d $586 \times 9 = 5274$

2 Lola, who is in Year 5, goes skating with her mum, dad, 8-year-old sister and
3-year-old brother. Below are the admission prices.

466
2578
   8

20624

Tickets	Morning 10:00 - 13:00	Afternoon 13:00 - 16:00	Evening 16:00 - 19:00	Weekend All day
Adults	£5.50	£6.00	£6.50	£7.50
Child (5-12)	£3.50	£4.00	£4.50	£5.00
Child (under 5)	£1.50	£2.50	£2.50	£3.00
Family ticket 2 adults and 3 children	£18.00	£21.00	£24.00	£27.00

12 20   22 50

a How much cheaper is it for the family to buy a family ticket than to
buy individual tickets during the afternoon?   22

£ **1.50**

b How much more expensive is it for an adult at the weekend than in
the morning during the week?

£ **2**

c How much would it cost to buy individual tickets for the family during
the week in the evening?

£ **24.50**

MASTERING - Addition, subtraction, multiplication and division   **Year 5**

3 Find the missing numbers in these calculations.

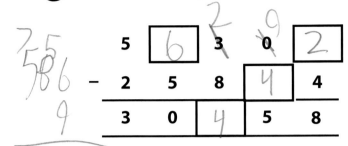

755
586
9
5274

5	6	3	0	2
– 2	5	8	4	4
3	0	4	5	8

2	7	9	6	5
+ 3	5	8	9	6
6	3	8	6	1

---

4 Bilal had some marbles.
He gave a **third** of his marbles to Ayesha. Ayesha gave a **third** of the marbles she received to Sadia.
Sadia kept **4** marbles and gave the remaining **2** to Lola.
How many marbles did Bilal have to begin with?

54

---

5 Tom and Bilal have raised **£425.40** between them.
Bilal raised **£54.30** more than Tom.
How much did each of the children raise?

Bilal = £ 239.85   **Tom = £** 185.55

239.85
185.55
425.40

---

6 Almost **5 million** Americans spend **42** hours each week playing computer games.
On average how many hours do they spend playing computer games every day?

6 **hours**

---

7 Charlie's book is **432** pages long.
If he reads **12** pages every day, how many days will it take him to finish reading the book?

036
12/432
36

36 **days**

---

8 **a** Lola has a pile of **2p** coins worth **£1**.
Each coin is **1.85 mm** thick.
How tall is Lola's pile of coins?

$$\begin{array}{r} 41.85 \\ 5 \\ \hline 9.25 \end{array}$$

92.5 mm

b Ayesha has a pile of **1p** coins measuring **15 cm**. Each coin is **1.5 mm** thick. How much is Ayesha's pile of **1p** coins worth?

£ ☐

c Tom has a pile of **5p** coins which is worth **£1.25**. The pile is **4.25 cm** tall. How thick is each of Tom's **5p** coins?

$$0.1 \quad 017$$
$$5\overline{)4.25} \qquad 1.7 \text{ mm}$$

d How much money do the **three** children have altogether?

£ **3.25**

9 There are **1136** children at South End High School.
587 children have packed lunches.
School lunches cost **£1.60** per child each day.
Mrs Smith collects the lunch money every Monday for the week.
So far, she has collected **£3600**.
How many children have not paid for their school lunch so far?

$$\begin{array}{r} 342.4 \\ 878.4 \\ 5 \\ \hline 43920 \end{array}$$

$$\begin{array}{r} 876 \\ 587 \\ \hline 3 \quad 54.9 \\ 3920 \\ 3600 \\ \hline 0792 \end{array}$$

$$\begin{array}{r} 259 \\ 599 \\ 1.6 \\ \hline 3294 \\ 5490 \\ \hline 8784 \end{array}$$

$$\begin{array}{r} 0495 \\ 1.6792 \, 0 \\ 64 \\ \hline 0080 \end{array}$$

99

10 Bilal accidentally tore a page out of his maths textbook.
Lola asked him what page it was.
Bilal said that the sum of the page numbers on the pages now facing each other was **351.**
What were the page numbers on the pages that Bilal tore out of the book?

175·5
2) 3 5 1 ¹ ¹⁰

| 175 | 176 |

---

11 Charlie and Lola were asked to buy some fruit for the after-school club.
They needed **18** apples, **10** bananas and **12** oranges.

apples 405	£1.35 per half dozen
plums	£1.55 per punnet of 8
bananas 220	£1.10 per bunch of 5
nectarines	84p per pair
orange 375	£1.25 per bag of 4

How much will it cost them?

405
220
375
————
1000

£ | 10 |

12 Bilal has decided to save some pocket money each month.
He saves **£1** in January, **£2** in February, **£3** in March, and so on.

How much money has Bilal saved after a year?

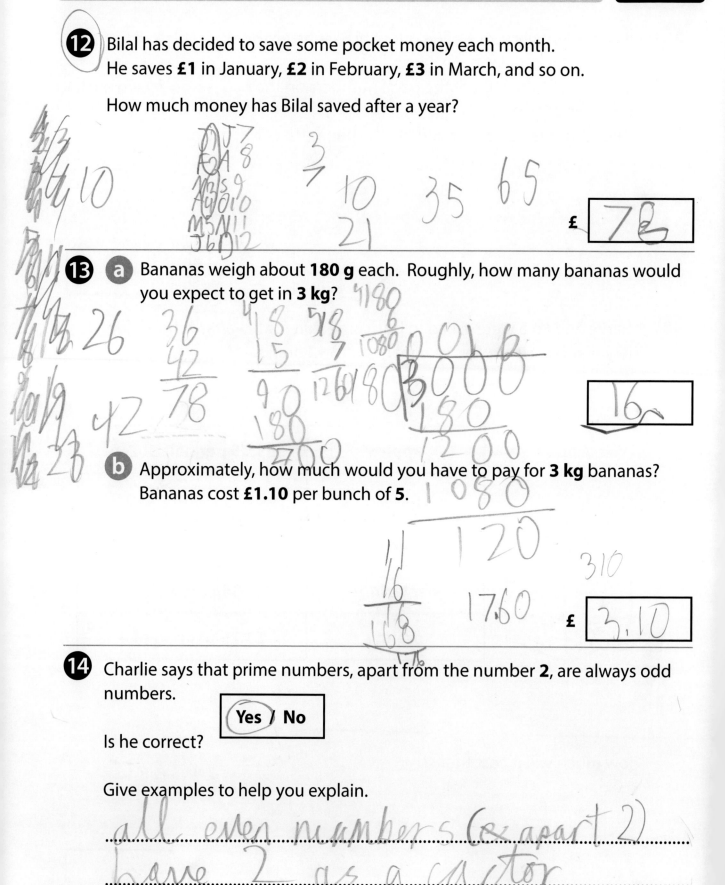

£ 78

13 **a** Bananas weigh about **180 g** each. Roughly, how many bananas would you expect to get in **3 kg**?

16

b Approximately, how much would you have to pay for **3 kg** bananas?
Bananas cost **£1.10** per bunch of **5**.

£ 3.10

14 Charlie says that prime numbers, apart from the number **2**, are always odd numbers.

Yes / No

Is he correct?

Give examples to help you explain.

all even numbers (except apart 2)
have 2 as a factor

15 Tom is thinking of **two** numbers.
The highest common factor of his numbers is **4**.
The lowest common multiple of his numbers is **48**.
One of the numbers is **12**.
What is the other number?

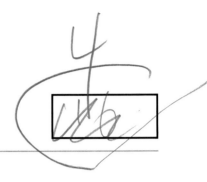

16 Complete this magic square using the numbers **0-15**.
The magic number is **30**.

12	2	1	15
7	9	10	4
14	5	6	8
0	14	13	3

17 **60** Strawberry Chews cost **£4.80**.

a How much do **20** Strawberry Chews cost?

£ 1·60

b How much do **100** Strawberry Chews cost?

£ 8.00

18 Mrs Palmer unpacks a box of exercise books at the start of term.
She puts them into piles of **30**, **one** pile for each class, and there are **10**
books left over.

There are **8** classes.

How many books were delivered to the school?

250

19 This is a multiplication pyramid.
You find the total in the box by finding the product of the **two** boxes
beneath it.

Complete this number pyramid.

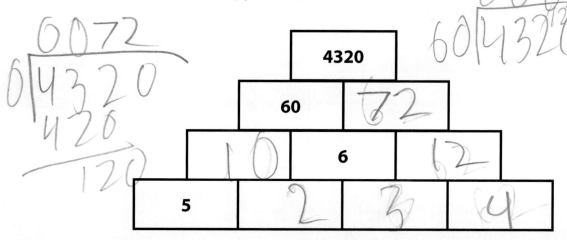

```
   0072
60|4320
   420
   120
```

```
  0062
60|4320
```

Pyramid:
- 4320
- 60 | 72
- 10 | 6 | 12
- 5 | 2 | 3 | 4

20 Lola says that the factors of every number can be organised into pairs.
This means that every number has an even number of factors.

Is Lola correct?   Yes / No   1, 2, 4   1, 2, 4

Use examples to help you explain.   12 - 1, 2, 3, 4, 6, 12

Square numbers   13 - 1, 13
less/extra

21 Use a formal written method to solve these calculations.

4756 + $\boxed{3941}$ = 8697

2647 = $\boxed{6181}$ − 3524

$4\overline{)9684}$ $\overset{2421}{}$ ÷ $\boxed{2421}$ = 4

$\boxed{1356}$ = $3\overline{)4068}$ ÷ 3

22 Write **four** number facts that this diagram shows.

7.6	
2.7	4.9

$2.7 + 4.9 = 7.6$ $7.6 - 4.9 = 2.7$

$4.9 + 2.7 = 7.6$ $7.6 - 2.7 = 4.9$

MASTERING - Addition, subtraction, multiplication and division **Year 5**

23 I know that **12 × 36 = 432**.

Find 2 other pairs of numbers that have a product of **432**?

$\boxed{24}$ × $\boxed{18}$ = **432** $\boxed{6}$ × $\boxed{72}$ = **432**

---

24 Tom says that if you add **two 2-digit** numbers together, the answer cannot be a **4-digit** number.

Is he correct? $\boxed{\text{Yes} / \text{No}}$

$\begin{array}{r} 99 \\ 99 \\ \hline 198 \end{array}$

Give examples to help you explain.

99 + 99 = 198 and 99 is the biggest two digit number

---

25 Sadia says if you multiply **two whole** numbers together (apart from **× 1**) the product is always bigger.

Is Sadia correct? $\boxed{\text{Yes} / \text{No}}$

Give examples to help you explain.

....

....

---

26 Using this number sentence as an example write down **2** different pairs of numbers with a total of **6**.

2.6 + 3.4 = 6

$\boxed{3}$ + $\boxed{3}$ = **6** $\boxed{2}$ + $\boxed{4}$ = **6**

    Name ....

27 Bilal starts with **386**.
He counts backwards in **4s**.
He says that if he keeps going for long enough, he will reach **0**.

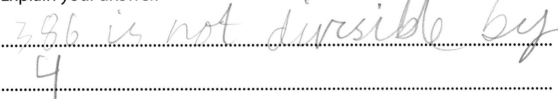

a Is Bilal correct?   Yes / No

Explain your answer.

386 is not divesible by
4

b What **3-digit** number could Bilal start from and count backwards in **3s** to reach **0**?

Checked   333

---

28 I know that **2.4 × 6.5 = 15.6**.

How can I use this number sentence to help me solve these calculations?
Find the answer and explain how you found it.

a 2.4 × 65 = 156

b 1.2 × 6.5 = 7.8

c 15.6 ÷ 2.4 = 6.5

d 156 ÷ 24 = 65

...d earns **£4500** in January, **£2325** in February and **£6975** in March.
He puts this money into his bank account where he already has **£27,800**.
He wants to decorate his lounge, so divides his total savings by **100** and
takes this amount out of the account.
How much money does Tom's dad have left in his bank account now?

£ 257522

30 The product of a **2-digit** and a **3-digit** number is **2400**.
What could the **two** numbers be?
Find two solutuions.

278800
278
27522

$$\boxed{24} \times \boxed{100} = 2400$$

$$\boxed{12} \times \boxed{200} = 2400$$

31 Sadia is thinking of a **2-digit** number.

When she divides it by **2**, the remainder is **1**.

When she divides it by **3**, the remainder is **1**.

When she divides it by **4**, the remainder is **1**.

When she divides it by **5**, the remainder is **1**.

When she divides it by **6**, the remainder is **1**.

What is the remainder when she divides it by **7**?

1

32 I know that if I want to multiply a number by **25**, I can multiply it by **100** and then divide by **4**. Use this fact to find these products.

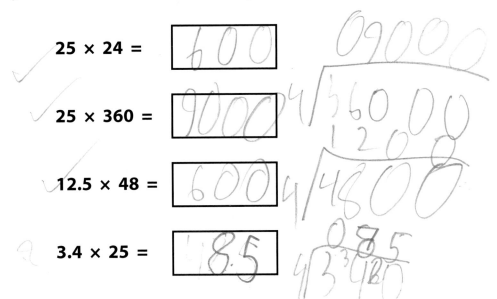

25 × 24 = ☐ 600

25 × 360 = ☐ 9000

12.5 × 48 = ☐ 600

3.4 × 25 = ☐ 85

---

33 Charlie wants to plant **16** apple trees.
He wants to arrange them in a rectangular array.

a List all the rectangles that Charlie could make with **16** apple trees.

 2 × 8   4 × 4   16 × 1

b Each apple tree needs **1** square metre.
Charlie wants to put a fence around his rectangle of apple trees.
He knows that each rectangle will need a different amount of fencing to surround it.
Fencing costs **£5** per metre.   80   170

What is the difference in price between the rectangle that needs the most fencing and the rectangle that needs the least fencing.

£ ☐ 90

---

34 Fill in the gaps to make the statements true. The first one has been done for you.

a | 12 | is a multiple of | 3 | and a factor of | 24 |

b | 10 | is a multiple of | 5 | and a factor of | 20 |

c | 24 | is a multiple of | 3 | and a factor of | 8 |

d | 30 | is a multiple of | 10 | and a factor of | 60 |

e | 3 | is a multiple of | 1 | and a factor of | 6 |

35 Lola goes to South Street Primary School.
Lola's mum and grandad also went to South Street Primary School.
This year, the teachers are organising a street party to celebrate the number of years that it has been open.
Lola knows that her mum's and grandad's ages are both square numbers and that the total of these numbers is the age of the school, which is also a square number.

How old is Lola's mum, grandad and South Street Primary School?

Lola's mum =

Lola's grandad =

South Street Primary School =

36 Ayesha bought some chocolates. x
She ate **3** of the chocolates before she went to school and gave her brother **2**. $x+5$
At school, she shared the rest of the chocolates with **6** of her friends.
Her **first** friend took **2** chocolates; the **second** friend took **4** chocolates; the **third** friend 6 chocolates; and so on. After the last friend had taken her chocolates, there were **none** left.

2 4 6 8 10 12   12 + 8 = 20
20 + 10 = 30
30 + 12 = 42

How many chocolates did Ayesha have at the beginning?

42 + 5
= 47

4 8
5 10
6 12

17 11 7

25

35   £ 47

NUMBER

Fractions
(including decimals and percentages)

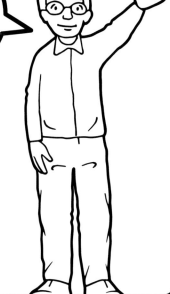

These are all about fractions, including decimals and percentages!

NUMBER - Fractions (including decimals and percentages)   **Year 5**

Compare and order fractions whose denominators are all multiples of the same number

1 Tom ate **two fifths** of a cake and Charlie ate **three tenths** of the cake. Who ate the most? Explain your answer. Tom as $\frac{2}{5} = \frac{4}{10}$ and $\frac{4}{10}$ is more than $\frac{3}{10}$

2 Lola eats **three quarters** of her packed lunch. Bilal eats $\frac{5}{8}$ of his lunch. Who has eaten the larger fraction of their lunch?

Lola

3 Ayesha uses $\frac{7}{10}$ of the felt tips for her drawing. Sadia, uses $\frac{3}{5}$ of the felt tips. Who uses the least felt tips for their drawing?

Sadia

4 Which fraction is bigger: $\frac{2}{3}$ or $\frac{3}{9}$? Explain your answer. $\frac{2}{3}$ as $\frac{3}{9}$

5 Tom, Charlie and Bilal have a pizza each. Tom eats $\frac{3}{4}$, Charlie eats $\frac{7}{12}$ and Bilal has $\frac{2}{3}$ of his pizza. Who eats the most pizza and who eats the least?

Tom eats the most and Charlie eats the least

6 Put these fractions in order of size, from largest to smallest.

$\frac{1}{2}$ $\frac{5}{6}$ $\frac{8}{12}$ $\frac{3}{4}$

$\frac{3}{4}, \frac{8}{12}$ $\frac{5}{6}, \frac{3}{4}, \frac{8}{12}, \frac{1}{2}$

7 Ayesha says, "$\frac{7}{8}$ is smaller than $\frac{12}{16}$." Is she correct? Explain your answer. No as $\frac{12}{16} \div 2 = \frac{6}{8}$ which is smaller than $\frac{7}{8}$

Recognise mixed numbers and improper fractions and convert from one form to the other

1 Tom works out how many **eighths** there are in $1\frac{5}{8}$. What should his answer be?

2 Is $\frac{11}{3}$ a mixed number or an improper fraction? Explain your answer.

3 Miss Whitehead writes down $\frac{5}{2}$ on the white board. She asks her class to write this improper fraction as a mixed number. What should they write?

4 There are **2** pizzas on the table. Ayesha is very hungry and eats $\frac{3}{2}$ of the pizzas. How much has Ayesha eaten as a mixed number? How much of the pizza is left over?

5 Bilal cuts his pies into **quarters**. He sells $5\frac{3}{4}$ of his meat and potato pies. How many **quarters** of pie has he sold?

6 Tom says, "**Twelve fifths** as a mixed number is $3\frac{2}{5}$." Is he correct? Explain your answer.

7 Convert $3\frac{4}{8}$ to an improper fraction.

8 Charlie ran $2\frac{3}{4}$ **km** and Sadia ran $3\frac{1}{4}$ **km**. How far did each person run as an improper fraction?

Recognise mixed numbers and improper fractions and convert from one form to the other

1 How many **eighths** are there in $1\frac{3}{8}$?

2 Is $\frac{8}{5}$ a mixed number or an improper fraction? Explain your answer.

3 Rufus the dog is $\frac{7}{2}$ years old. How old is Rufus, as a mixed number?

4 Lola is making lemonade. She uses $\frac{17}{3}$ lemons. How many lemons has she used as a mixed number? How many whole lemons has she used?

5 The baker sells $5\frac{1}{3}$ of his custard pies. (He cuts them into **thirds**.) How many **thirds** of custard pie does he sell?

6 Bilal says, "$\frac{11}{6}$ as a mixed number is $1\frac{6}{6}$." Is he correct?
Explain your answer.

7 Convert $\frac{22}{8}$ to a mixed number.

8 Ayesha split each orange into **10** segments. She ate **4** full oranges and **7** extra segments. What fraction of the orange segments did Ayesha eat altogether? Show your answer as an improper fraction.

Name ....

Add fractions with the same denominator

1 Sadia adds $\frac{1}{4}$ and $\frac{2}{4}$. What should her answer be?

2 Lola eats $\frac{1}{5}$ of her grapes at break time. She then eats another $\frac{3}{5}$ at lunch. What fraction of her grapes has she eaten altogether?

3 Bilal is trying to work out what $\frac{2}{7}$ add **three sevenths** would be. What fraction should he write down?

4 Tom eats $\frac{4}{8}$ of his pizza for his dinner. He then eats another $\frac{3}{8}$ the next day for his lunch. How much of the pizza has he eaten in total?

5 Sadia walks $\frac{4}{12}$ around the perimeter of the lake. After a rest, she then walks a further $\frac{5}{12}$. How much of the perimeter of the lake has Sadia walked?

6 Add together **four fifteenths** and $\frac{8}{15}$.

7 Charlie spends $\frac{4}{16}$ of his money on a t-shirt, $\frac{2}{16}$ on a belt and $\frac{7}{16}$ on a new game. What fraction of his money has Charlie spent?

8 Lola adds these fractions:

$$\frac{3}{22} \qquad \frac{5}{22} \qquad \frac{7}{22} \qquad \frac{2}{22}$$

What should her answer be?

89   Name ....

NUMBER - Fractions (including decimals and percentages)   Year 5

Subtract fractions with the same denominator

1 Tom subtracts $\frac{2}{4}$ from $\frac{3}{4}$. What should his answer be?

$\frac{1}{4}$

2 $\frac{4}{5}$ of a class like pizza. $\frac{2}{5}$ like only ham pizza. The rest like any pizza. What fraction of the class like any pizza?

$\frac{2}{5}$

3 Charlie's mum puts $\frac{6}{8}$ of the mince pies on the table. Charlie eats $\frac{3}{8}$. What fraction of the pies are left on the table?

$\frac{3}{8}$

4 Mrs Penrose puts $\frac{7}{9}$ of the paints on her desk. She uses $\frac{4}{9}$ of the paints in her lesson. What fraction of the paints are left on her desk?

$\frac{3}{9} = \frac{1}{3}$

5 Find the difference between $\frac{5}{12}$ and $\frac{10}{12}$. Write your answer as a fraction.

$\frac{5}{10} \frac{5}{12}$

6 Ayesha has worked out that $\frac{9}{10} - \frac{3}{10} = \frac{5}{10}$. Is she correct? Explain your answer. No as

No as 9-3=6 not 5

7 Bilal's brother has $\frac{11}{18}$ of his wage left. He spends another $\frac{5}{18}$ of his wage on a trip to the zoo. What fraction of his wage does he have left now?

$\frac{6}{18} = \frac{1}{3}$

8 Subtract $\frac{12}{30}$ from $\frac{23}{30}$.

$\frac{11}{30}$

✓ checked

NUMBER - Fractions (including decimals and percentages)   **Year 5**

Add and subtract fractions with the same denominator

1 Tom adds $\frac{1}{6}$ and $\frac{2}{6}$. What should his answer be?

$3 = 1.$
$\frac{}{6} = \frac{}{2}$

2 $\frac{4}{7}$ of the class cycle to school. $\frac{2}{7}$ cycle less than a mile. What fraction of the class cycle a mile or more?

$\frac{2}{7}$

3 Ayesha is trying to work out what $\frac{3}{9}$ add **two ninths** is. What fraction should she write?

$\frac{5}{9}$

4 Mr Wood asks his class to find the difference between $\frac{2}{9}$ and $\frac{7}{9}$. What is the answer, as a fraction?

$\frac{5}{9}$

checked

5 Sadia sprints around $\frac{3}{8}$ of the park perimeter. She then jogs a further $\frac{2}{8}$. What fraction of the perimeter has she travelled?

$\frac{5}{8}$

6 Bilal has $\frac{7}{12}$ of his pocket money left. He spends a further $\frac{1}{12}$ on a book and $\frac{5}{12}$ on the cinema. What fraction of his pocket money does Bilal have left now?

$\frac{1}{12}$

7 $\frac{1}{2}$ of the football crowd support the blue team. $\frac{7}{18}$ of the crowd who support the blue team are women. What fraction of the blue team supporters are men?

$\frac{}{18} = \frac{}{2} \quad \frac{11}{18}$

NUMBER - Fractions (including decimals and percentages)   **Year 5**

Add fractions with denominators that are multiples of the same number

1 Charlie adds $\frac{1}{4}$ and $\frac{1}{2}$. What should his answer be?

$\frac{3}{4}$

2 Sadia eats $\frac{1}{3}$ of a cheese pizza for her lunch and $\frac{3}{6}$ for her tea. How much pizza has she eaten altogether?

$\frac{5}{6}$

3 Mr Harrison asks his class to work out the answer to $\frac{3}{10}$ plus $\frac{2}{5}$. What answer should they give?

$\frac{7}{10}$

4 What is the total of **three sixths** and $\frac{3}{12}$?

$\frac{9}{12}$ Checked

5 Bilal ran around $\frac{2}{4}$ of the track. He then ran around another $\frac{3}{8}$. How much of the track did he run around altogether, as a fraction?

$\frac{7}{8}$

6 At the cake sale, $\frac{3}{5}$ of the cakes are sold at break time and $\frac{4}{10}$ are sold at lunch. What fraction of the cakes are sold altogether?

$1 = \frac{5}{5} = \frac{10}{10}$ $\frac{3}{5} + \frac{4}{10}$

7 Ayesha spends $\frac{1}{2}$ of her money on a dress and $\frac{4}{12}$ on a bracelet. She then spends $\frac{1}{12}$ on a book. How much money has she spent? Write your answer as a fraction.

$\frac{11}{12}$

8 Lola adds these fractions together: $\frac{3}{6}$, $\frac{1}{4}$, $\frac{2}{12}$. What should her answer be?

$\frac{11}{12}$

NUMBER - Fractions (including decimals and percentages)   **Year 5**

Subtract fractions with denominators that are multiples of the same number

1 Lola has $\frac{3}{4}$ of an apple pie left. She eats another $\frac{1}{2}$ for her lunch. How much apple pie is left?

$\frac{1}{4}$

2 $\frac{5}{6}$ of the class can swim. $\frac{2}{3}$ can swim only breaststroke. What fraction of the class can swim other strokes?

$\frac{1}{6}$

3 Charlie wants to work out the difference between $\frac{6}{8}$ and $\frac{1}{4}$. What should his answer be?

$\frac{4}{8} = \frac{1}{2}$

4 $\frac{8}{10}$ of the bean bags are thrown more than **10 metres**. $\frac{3}{5}$ of the bean bags are thrown more than **15 metres**. What fraction of the bean bags are thrown between **10 metres** and **15 metres**?

$\frac{2}{10} = \frac{1}{5}$

5 There are $\frac{10}{12}$ of the pies left in the bakery. They sell another $\frac{4}{6}$. What fraction of pies are left to sell?

$\frac{2}{12} = 1$

6 Tom has worked out that $\frac{7}{8} - \frac{2}{4} = \frac{4}{8}$. Is he correct? Explain your answer.

No as $\frac{2}{4} = \frac{4}{8}$, $\frac{7}{8} - \frac{4}{8} = \frac{3}{8}$ Checked

7 Subtract $\frac{3}{4}$ from $\frac{5}{6}$.

$\frac{1}{12}$

8 Bilal has saved $\frac{10}{12}$ of his spending money. He spends $\frac{1}{3}$ to watch a football match and $\frac{2}{6}$ on a scarf. What fraction of his spending money does he have left?

$\frac{2}{12}$

Add and subtract fractions with denominators that are multiples of the same number

1 Sadia adds $\frac{2}{4}$ and $\frac{1}{2}$. What is her answer?

2 $\frac{4}{5}$ of the pupils are in assembly. $\frac{5}{10}$ of the pupils are in assembly and sitting on chairs. What fraction of the pupils are in assembly and not sitting on chairs?

3 Lola is trying to work out what $\frac{3}{6}$ plus **two thirds** is. What fraction should she write?

4 The frog swims $\frac{3}{4}$ of the pond's length. It then swims another $\frac{2}{8}$. What fraction of the pond's length did the frog swim in total?

5 Ayesha eats $\frac{2}{8}$ of her raisins at breakfast and $\frac{1}{2}$ at lunch. What fraction of her raisins has she eaten?

6 Sadia ran $\frac{7}{9}$ of the marathon. Tom ran $\frac{15}{18}$. What is the difference between the distances they ran, as a fraction?

7 Bilal subtracts $\frac{3}{6}$ from $\frac{3}{4}$. What should his answer be?

8 Charlie adds $\frac{5}{6}$ to $\frac{6}{9}$. What should his answer be?

Multiply proper fractions and mixed numbers by whole numbers

1 There are **2** oranges in a bowl. Tom and his sister eat $\frac{1}{2}$ an orange each. How many oranges have they eaten in total?

2 **Four** friends are having a sleepover. They eat $\frac{1}{2}$ of a chocolate bar each. How many bars of chocolate have they eaten in total?

3 Mum buys pizzas for her **five** children. They each eat $\frac{1}{2}$ of a pizza. How many pizzas did they eat altogether?

4 Charlie is confused about how to multiply fractions. He is trying to work out $1\frac{3}{4} \times 2$. Show Charlie how to work this out.

5 Bilal is trying to work out the answer to $2\frac{4}{8} \times 4$. What should his answer be?

6 Miss Andrews asks Year 5 to work out this calculation:

$$7\frac{5}{6} \times 8$$

What should the answer be?

7 There are **5** people in the Brown family and each person has a pie on their dinner plate. They each eat **three quarters** of their pie. How many pies have been eaten? Write your answer as a mixed number.

Read and write decimal numbers as fractions

1 Tom has been trying to find a fraction which is the same as **0.5**. Which fraction could he choose?

2 Which decimal number is the same as $\frac{1}{4}$?

3 Which of the fractions below is the equivalent to **0.3**?

$$\frac{3}{10} \qquad \frac{4}{10} \qquad \frac{33}{10}$$

4 Sadia has worked out that **0.8** as a fraction is $\frac{8}{10}$. Is she correct? Explain your answer.

5 Lola is working with Sadia and says that $\frac{80}{100}$ is also equivalent to **0.8**. Is she correct? Explain your answer.

6 Grandad can't remember what fraction is the same as **0.75**. What is the answer?

7 Can you work out what fraction is the same as **0.25**?

8 Bilal is trying to work out which improper fraction is equal to **1.2**. What should his answer be?

Recognise and use thousandths and relate them to tenths, hundredths and decimal equivalents

1 A survey found that **123** people out of **1000** people didn't like chocolate. Show this as a fraction.

2 Mr Fletcher asks Year 5 to work out the equivalent fraction to **0.431**. What answer should they give?

3 What is the decimal equivalent of $\frac{75}{1000}$?

4 Ayesha's teacher asked her how many **tenths** were equivalent to $\frac{200}{1000}$. What should her answer be?

5 How many **hundredths** are equivalent to $\frac{980}{1000}$?

6 **363** people out of a **thousand** prefer having a bath to a shower. Show this as a fraction and as a decimal number.

7 Bilal has **1000** cards. **320** of them are football cards. How many **hundredths** of his cards are football cards?

8 Tom states, "$\frac{1}{8}$ is equivalent to **0.125**." Is he correct? Explain your answer.

Round decimals with two decimal places to the nearest whole number and one decimal place

1 Round **2.22** to the nearest whole number.

2 Ayesha wants to round **4.91** to the nearest whole number. What answer should she give?

3 Mr Rajan asks his class to round **6.34** to **one** decimal place. What number should they say?

4 Round **41.86** to the nearest whole number.

5 Bilal says, "**64.64** rounded to **one** decimal place is **64.7**." Is he correct? Explain your answer.

6 Charlie rounds **143.49** to the nearest whole number. What number should he write down?

7 Tom has **£22.33** in his wallet. How much money does Tom have to the nearest **pound**?

8 Bilal has **£78.66**. To the nearest **ten pence**, how much money does Bilal have?

Read, write and order numbers with up to three decimal places

1   Mrs Yeung writes the numbers **0.35** and **0.53** on the whiteboard. Which number is smaller?

2   In the shop, a chocolate bar is **£1.63**, a choc ice is **£1.82** and a toffee bar is **£1.75**. Put the items in order, starting with the most expensive.

3   It takes Ayesha **18.43 seconds** to run 100 metres. Tom's time is **18.438 seconds**. Who is faster? Explain your answer.

4   Which is longer: **4.635 m** or **4.653 m**?

5   In a long jump competition, Sadia jumps **3.56 m**, **3.562 m** and **3.089 m**. Rank her jumps, starting with the shortest.

6   Put these decimals in order of size starting with the smallest:

       43.62   **43.764**   **43.72**   **43.671**

7   Counting up in **thousandths**, what is the next number after **0.923**?

8   Which of these decimals is the smallest: **59.669**, **59.664**, **59.71** or **59.69**? Explain your answer.

Solve problems involving numbers with up to three decimal places

1 Bilal divided **12.4** by **4**. What should his answer have been?

2 Charlie calculated **12.45** plus **18.36**. Write down what you think his answer should have been.

3 Lola's average computer game score is **18.543**. Tom's average score is **12.42**. Find the difference in their scores.

4 Ayesha runs **3.654 kilometres** on Monday and **6.344 kilometres** on Tuesday. How far does she run in total?

5 Tom ran twice as far as Ayesha ran. Ayesha ran **6.344 km**. How far did Tom run?

6 Bilal says a number divided by **2** is **4.651**. What is the number?

7 Charlie finds the total of **4.01** and **8.493**. He then subtracts **2.38**. What should his answer be?

8 A shop sells **46.3%** of its stock on Saturday. On Monday, it sells **38.456%**. What percentage is there left to sell?

Recognise the percent symbol and solve percentage problems

1 A designer jacket usually costs **£80**. In a sale, all items are reduced by **25%**. How much does the jacket cost in the sale?

2 A tortoise race track was **400 cm** long. Speedy, the tortoise, completed only **20%** of the track. How many **centimetres** did he travel?

3 There are **200** pupils at Blackrock Primary School. **48%** are girls. How many are boys?

4 Sadia tossed a coin **20** times. It landed on heads **15** times. What percentage of the tosses did the coin land on heads?

5 In a quiz, The Brainy Boffins scored **75** correct answers out of **80**. In the same quiz, The Potty Professors scored **90%** correct answers. Which team won the quiz?

6 Bilal ate **50%** of a pizza. Tom ate **50%** of what was left. What percentage of the original pizza was left?

7 **300** people were asked if they preferred watching TV or listening to music. **60%** said they preferred watching TV. How many people preferred listening to music?

8 A dress costing **£26** was reduced by **25%** in a sale. Lola had **£24**. Did she have enough money to buy the dress? Explain your answer.

NUMBER - Fractions (including decimals and percentages)   Year 5

Write percentages as a fraction with the denominator hundred, and as a decimal

1 Lola eats **87%** of her bag of grapes at break time. What fraction of grapes has she eaten?

$\frac{87}{100}$

2 At Summergill School, **53%** of students are female. How many students are female, as a fraction?

$\frac{53}{100}$

3 Sadia achieved **67%** in her maths test. What fraction did she get correct in her maths test?

$\frac{67}{100}$

4 Charlie is buying a new t-shirt in the sale which has **75%** off. Charlie knows that the percentage as a fraction is $\frac{75}{100}$. What would the percentage be as a decimal number? 0.75

Checked

5 Write **52%** as a decimal number.

0.52 ✓

6 In the football stadium, **69%** of the fans are supporting Hillingham. What fraction of the fans are supporting Hillingham?

$\frac{69}{100}$

7 The rest of the fans (from the question above) are Whelsea fans. What percentage of fans are Whelsea fans? What would this be as a decimal?

0.31

8 What would be the equivalent decimal number to **89.5%**?

0.895

NUMBER - Fractions (including decimals and percentages)   **Year 5**

Solve problems which require knowing percentage and decimal equivalents

1 Mrs Walker writes down **50%** on the board. What is the decimal equivalent?

0.5

2 Work out the percentage equivalent to **0.25**.

25%

3 **0.75** of the children are girls. What percentage of the children are boys?

25%

4 Sadia is trying to work out the difference between **60%** and **40%**. What is her answer as a decimal number?

0.2 ✓ Checked

5 Ayesha thought that **0.35** was bigger than **35%**. Is she correct? Explain your answer. *She is wrong as 0.35 = 35%*

6 **90%** of the school children attended the school disco. How many children did not attend the disco? Write your answer as a decimal number.

0.1

7 What is the percentage equivalent to **0.125**?

12.5%

8 **48%** of the children at Rosetree Primary School are girls. What proportion are boys? Write your answer as a decimal number.

0.52

NUMBER - Fractions (including decimals and percentages)   Year 5

Solve problems involving fractions

1 Tom was asked by his teacher to explain how he could find **one sixth** of a number. What should he say? He should say "To find one sixth of a number you divide the number by 6."

2 Sadia wanted to know which keys on the calculator she should press to work out **one eighth** of **96**. Write down, in the correct order, the keys she should press to find the answer.

96 ÷ 8 =

3 Ayesha worked out that $\frac{1}{6}$ of a number was **12**. What was the number that she started with?

72

4 Lola had collected **172** pictures of ponies from magazines. Ayesha had collected **three quarters** of this amount. How many pictures had Ayesha collected?

043
4|172

43
3
129

129

Checked

5 Mr Sikora is **33**. His brother is $\frac{2}{3}$ of his age. How old is Mr Sikora's brother?

22

6 Bilal has **six hundred** stickers. He put **four fifths** of them into his sticker album. How many stickers does he still need to put in?

480

120

7 Mr Button saved up **£1100** to buy his first car. He realised that he would need to spend $\frac{2}{5}$ of this amount on the insurance. How much would he have left to buy the car?

120
5|600

£660

0220
5|1100

W17   7-1-24

Solve problems involving fractions

1 What is **one fifth** of **twenty**?

4

2 Bilal had **32** pages to read in his reading book. Lola had $\frac{3}{4}$ of this amount still to read. How many pages did Lola still have to read?

24

3 **Five sixths** of a group of **72** children like cheese. How many children like cheese?

60

4 Tom wanted to raise some money for Children in Need. He did a sponsored swim and raised **two fifths** of his target of **£60**. How much did he raise in the sponsored swim?

£24

5 What is **one hundredth** of **1 litre**? Give your answer in **millilitres**.

10 ml   ✓   *Checked*

6 Which is more: $\frac{3}{5}$ of **£70** or **one half** of **£80**?

$\frac{3}{5}$ of £70

7 Lola sleeps for **10** hours every day. What fraction of the day is she awake for? Give your answer in the lowest terms.

$\frac{14}{24} = \frac{7}{12}$

8 Out of **36** children, $\frac{1}{9}$ like swimming best, $\frac{1}{3}$ like netball and $\frac{1}{9}$ like football. The rest like running best. How many children like running best?

16

It may be appropriate for children to use exercise books or paper to record their answers, working out or explanations.

MASTERING

Fractions
(including decimals and percentages)

Prepare to use all your thinking power!

1 Make each number sentence correct using **<, >, =**

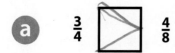

a $\frac{3}{4}$ ▢ $\frac{4}{8}$ $\frac{1}{4}$ ▢ $\frac{4}{16}$ $\frac{8}{12}$ ▢ $\frac{3}{4}$

b $1\frac{1}{4}$ ▢ 1.2 2.4 ▢ $2\frac{2}{5}$ $\frac{7}{8}$ ▢ $\frac{15}{16}$

c $\frac{7}{3}$ ▢ $2\frac{2}{3}$ $4\frac{1}{2}$ ▢ $\frac{33}{8}$ 5.75 ▢ $5\frac{3}{4}$

---

2 Draw lines to match each decimal number to its equivalent fraction.

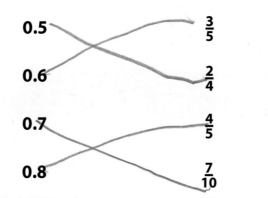

0.5 $\frac{3}{5}$

0.6 $\frac{2}{4}$

0.7 $\frac{4}{5}$

0.8 $\frac{7}{10}$

---

3 Mark and label on this number line where you estimate that these fractions should be positioned:

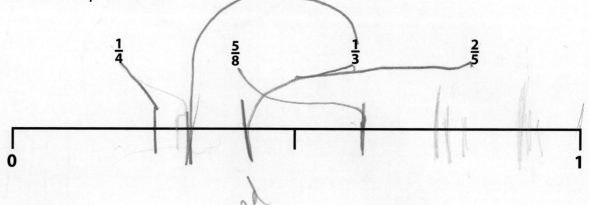

$\frac{1}{4}$ $\frac{5}{8}$ $\frac{1}{3}$ $\frac{2}{5}$

0   1

---

4 Tom wants to buy a computer game that costs **£35**.
In Shop **A**, the game has $\frac{1}{7}$ off.
In Shop **B** it had **15%** off.

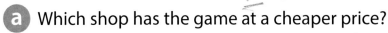

a Which shop has the game at a cheaper price?

shop   A B

b How much cheaper is it?

2·5   **p**

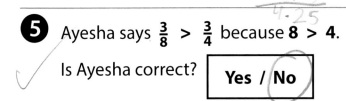

5 Ayesha says $\frac{3}{8}$ > $\frac{3}{4}$ because **8 > 4**.

Is Ayesha correct?   **Yes / No**

Explain your answer.

$\frac{3}{4} \times 2 = \frac{6}{8}$, $\frac{6}{8}$ is bigger than $\frac{3}{8}$

6 Choose numbers for each numerator to make this number sentence true.
Find **3** solutions.

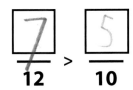   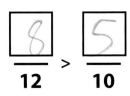   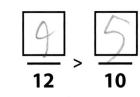

$$\frac{7}{12} > \frac{5}{10} \qquad \frac{8}{12} > \frac{5}{10} \qquad \frac{9}{12} > \frac{5}{10}$$

7 The answer to a calculation is **$2\frac{3}{5}$**.
What might the calculation be?
Use two mixed numbers.

$1\frac{1}{5} + 1\frac{2}{5}$

8 **a** Which is closer to **1**: $\frac{5}{6}$ or $\frac{21}{24}$?

Explain your answer.

$\frac{21}{24}$ as $\frac{5}{6} \times 4 = \frac{20}{24}$

b Write a fraction that is even closer to **1**?

$\frac{23}{24}$

9 **a** Find $\frac{40}{100}$ of **200**.

80

b Find $\frac{4}{10}$ of **200**.

80

c What do you notice?

they are the same

10 Lola loves cheese and marmite pizza.
On Monday, Lola ate the amount of pizza shown shaded on A.
On Tuesday, Lola ate the amount of pizza shown shaded on B.

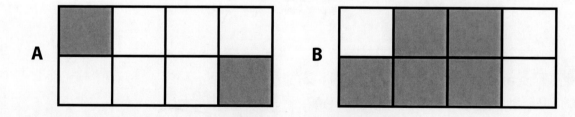

A   B

How much more, as a fraction, did Lola eat on Tuesday?

$\frac{3}{8}$

11 Tom and Sadia are watching a film and eating chocolate.
They have two identical bars of chocolate each.
Tom ate $1\frac{3}{8}$ of his chocolate.
Sadia ate $\frac{5}{4}$ of her chocolate.
Shade the diagrams below to show how much each child ate.

W47

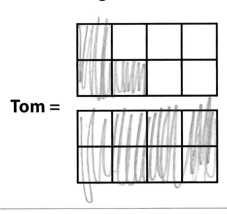

Tom =   Sadia =

12 Fill in this table to show how much of this diagram has been shaded.
One has been done for you.

	Black	Grey	White
Hundredths	10	40	50
Tenths	1	4	5
Decimal	0.1	0.4	0.5
Percentage	10%	40%	50%

checked

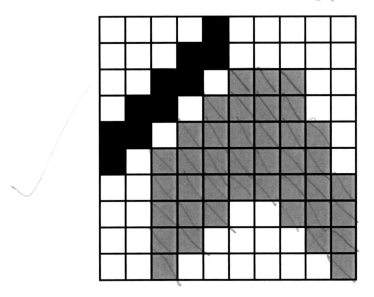

13 Using the numbers **2** and **3** only once, make this sum have the smallest possible answer.

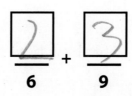

$$\frac{2}{6} + \frac{3}{9}$$

14 Write the numbers in order of size starting with the smallest:

a

4.05	5.4	4.5	5.50	5.04
4.05	4.5	5.04	5.4	5.50

b

7.07	7.7	7.67	7.6	6.7
6.7	7.07	7.6	7.07	7.7

15 Using the numbers **2**, **3**, **4** and **5** only once, make this sum have the largest possible answer.

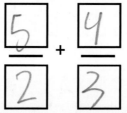

$$\frac{5}{2} + \frac{4}{3}$$

16 Circle the calculation which gives the largest answer:

30% of **900**   or   (35% of **800**?)

Explain how you know.

35% of 800 ....

....

17 Bilal gives his friends a glass of cola.
Each glass can hold $\frac{3}{5}$ of a bottle of cola.
Bilal has **four** bottles of cola.

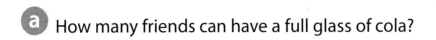

a How many friends can have a full glass of cola?

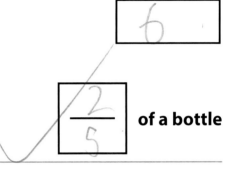

6

b How much cola is left over?

$\dfrac{2}{5}$ **of a bottle**

18 Ayesha and Lola go out shopping.
Ayesha spends $\frac{3}{4}$ of her money.
Lola spends **70%** of her money.

a Can you tell who has spent the most money?   **Yes / No**

Explain your answer.

We dont know how much
money they started with

b Give examples of amounts of money that they could have so that:

Ayesha spent more:

Ayesha = £ 20   /5   **Lola = £** 10

Lola spent more:

Ayesha = £ 10   **Lola = £** 10

19 How many fractions with a denominator of **100** can you find that have a value smaller than **0.1**?

$\frac{0}{100}$ $\frac{1}{100}$ $\frac{2}{100}$ $\frac{3}{100}$ $\frac{4}{100}$ $\frac{5}{100}$ $\frac{6}{100}$ $\frac{7}{100}$

$\frac{8}{100}$ $\frac{9}{100}$ (10)

20 **750,000** people went to watch a football team during one year.
The table shows how many people went each month.
Complete the table by filling in the missing totals and percentages.

	Totals	Percentage
January	75,000	10%
February	37500	5%
March	45,000	6%
April	52500	7%
May	60,000	8%
June	67,500	9%
July	22500	3%
August	30,000	4%
September	90,000	12%
October	82500	11%
November	112500	15%
December	75000	10%

21 Write the next number in these sequences:

0.067   0.068   0.069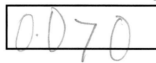

1.173   1.183   1.193

22 Write a number that is:

a exactly halfway between **3.45** and **3.46**.

b bigger than **0.76** and smaller than **0.77**.

0.765

23 $\frac{1}{5}$ of this rectangle is shaded.

The same rectangle is used in these diagrams.
What fraction of these diagrams is shaded?

a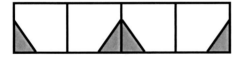

$\frac{4}{20}$

b

$\frac{1}{15}$

24 Circle the correct answer:

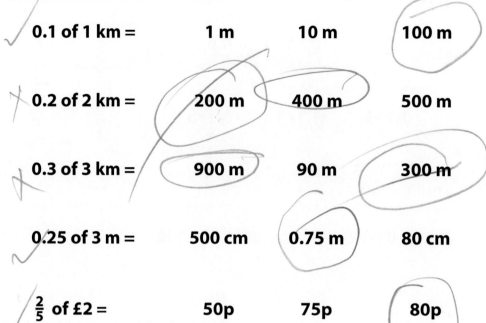

0.1 of 1 km =	1 m	10 m	(100 m)
0.2 of 2 km =	200 m	400 m	500 m
0.3 of 3 km =	900 m	90 m	300 m
0.25 of 3 m =	500 cm	0.75 m	80 cm
$\frac{2}{5}$ of £2 =	50p	75p	80p

25 Part of this number line is shaded.

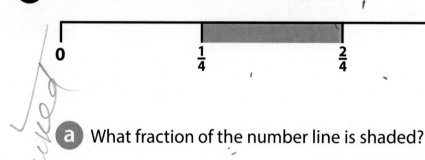

0    $\frac{1}{4}$ $\frac{2}{4}$ $\frac{3}{4}$ 1

a What fraction of the number line is shaded?

 $\frac{1}{4}$

b Write this fraction as a decimal.

 0.25

26 Give 3 examples of a fraction that is more than $\frac{7}{8}$.

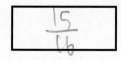 $\frac{15}{16}$ 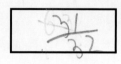 $\frac{31}{32}$ 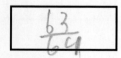 $\frac{63}{64}$

114   Name ....

27 Order these fractions from smallest to largest:

$\frac{2}{5}$ $\frac{3}{10}$ checked $\frac{4}{20}$

$\dfrac{4}{20}$ $\dfrac{3}{10}$ $\dfrac{2}{5}$

28 A supermarket car park has spaces for **1000** cars.

1% of the spaces are for electric cars, $\frac{1}{25}$ are for disabled drivers, $\frac{1}{20}$ are for parent and child drivers and the rest are free for anyone to use.

On Tuesday, the car park was $\frac{3}{4}$ full.

$\frac{1}{2}$ of the electric car spaces were used, $\frac{1}{4}$ of the disabled spaces and $\frac{4}{5}$ of the parent and child.

How many cars were there in each of the areas of the car park on Tuesday?

$25\overline{)1000}$ 0040
750

electric cars	5
disabled drivers	10
parent and child	40
other cars	695

29 **a** What needs to be added to **9.45** to give **9.69**?

0.24

b What needs to be added to **5.965** to give **6**?

0.035

30 Put the correct symbol in the box: <, >, =

3.60 [>] 3.06

15.425 [>] 14.425

WY7 5-8-24 Checked ✓

---

31 Give **two** examples of decimals which, when rounded to **1 decimal place**, are **13.4**.

✓

13.44   13.39

---

32 Here is a pattern on a grid.

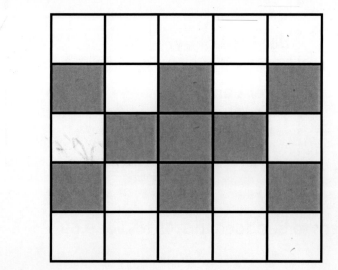

What percentage of the grid is shaded?

$\frac{9}{25}$

36 %

33 Here is a grid of **25** squares.

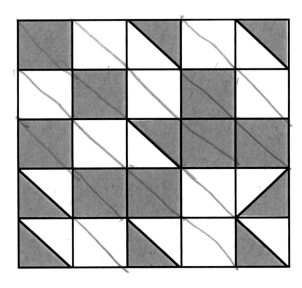

What percentage of the grid is shaded?

48 %

34 Circle the odd one out in the groups below. Explain why it is different.

a $\frac{6}{10}$ $\frac{12}{20}$ $\frac{3}{5}$ $\frac{12}{20}$ $\frac{18}{20}$ $\frac{9}{15}$ $\frac{12}{20}$

All of the others are equal
to $\frac{12}{20}$

b $\frac{30}{100}$ $\frac{3}{10}$ $\frac{6}{20}$ $\frac{3}{9}$

the others are equal to
$\frac{30}{100}$

35 What is $\frac{4}{5}$ as a percentage?

N 47
5-8-24

80 %

36 What is $\frac{22}{40}$ as a percentage?

Checked

55 %

37 Circle the percentage that is equal to $\frac{3}{5}$.

50%   (60%)   70%   80%   90%

38 Complete the pattern:

$\frac{71}{100}$	$\frac{81}{100}$	$\frac{91}{100}$	$\frac{101}{100}$
0.71	0.81	0.91	1.01

0.840 1
0.800 3
0.084 4
0.820 2
0.080 5

39 Put these numbers in order starting with the largest:

0.84	$\frac{8}{10}$	0.084	82%	$\frac{8}{100}$
0.84	82%	$\frac{8}{10}$	0.084	0.08

Explain how you can tell which is the largest and which is the smallest.

....

....

40 Lola and Charlie share **£56** so that Lola gets $\frac{2}{5}$ of what Charlie gets. How much do they each get?

2 of 56
8
22.40

QE

Lola = £ [ ]   **Charlie = £** [ ]

41 $\frac{7}{5}$ of **90** is **126**.
Complete the calculation below.

$$\frac{27}{20} \text{ of } 60 = 81$$

42 Add a suitable numerator or denominator to make each fraction equivalent to $\frac{2}{5}$.

$$\frac{4}{10} \qquad \frac{12}{20} \qquad \frac{6}{15}$$

43 Write down a number with **3** decimal places which, when multiplied by **100** gives an answer between **46** and **49**.

0.473

44 I divide a number by **1000** and the answer is **4.567**.
What number did I start with?

4576

WHT
5-8-24

45 Circle the larger fraction: $\frac{1}{3}$ $\frac{5}{15}$ $\left(\frac{2}{5}\right)$ $\frac{6}{15}$

Explain how you know.

I put them both into 15ths

46 Sadia makes a chocolate crispy cake.
It weighs **180 g**.
25% of the weight is chocolate chips, $\frac{1}{6}$ of the weight is rice pops.
The rest is sugar.
How many grams of each ingredient does Sadia use?

chocolate chips | 45 | g

rice pops | 30 | g

sugar | 105 | g

✓

47 A ball is dropped from a height of **100 cm** and bounces up to **60%** of the height from which it was dropped.
The second bounce is **40%** of the height of the first bounce. How high is the second bounce?

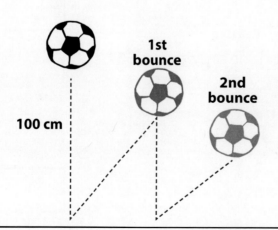

100 cm

1st bounce

2nd bounce

24
40 cm

W47
5-8-24

Checked

48 Tom thinks of a number.
He adds $\frac{1}{3}$ of the number to $\frac{1}{4}$ of the number.
The result is **35**.

What was the number Tom first thought of?

60

49 Circle the correct value of the **9** in the number **4.796**.

$\frac{9}{10}$ $\frac{9}{100}$ 9 $\frac{9}{1000}$

50 Circle **two** of these numbers which, when multiplied together, have the answer closest to **45**?

9.3   7.6   4.9   8.4

51 What fraction of **£4** is **55p**?

$\frac{55}{400}$ $\frac{11}{80}$ $\frac{11}{80}$

52 Put a circle around the smallest number:

0.506   0.65   5.06   6.05   0.605   0.56

53 Circle **all** the numbers that are smaller than **2.3**:

2.134   2.4   2.298   2.301   2.333

54 Round each decimal to the nearest whole number:

5.53	7.05	8.65
6	7	9

55 Divide **6.7** by **100**.

0.067

56 The peel of a banana weighs $\frac{3}{8}$ of the total weight of the banana.
If you buy **3 kg** of bananas at **96p** per **kg**, how much are you paying for the banana peel?

How much are you paying for the banana itself?

banana peel = £ 1.08

banana = £ 1.80

$\frac{3}{8} \times$

$\frac{2.88}{100}$

$3 \times \frac{3}{8}$

57 Circle the number closest in value to 1.

0.96   1.03   0.6   1.1

58 Charlie makes a fraction using **two** number cards.
He says, "My fraction is equivalent to $\frac{1}{3}$."
One of the number cards is **9**.
What could Charlie's fraction be?
Give both possible answers.

$\frac{9}{27}$ or $\frac{3}{9}$

59 Write these fractions in order of size, starting with the smallest.

$\frac{5}{8}$ *15* $\frac{1}{2}$ $\frac{2}{3}$ *16* $\frac{7}{12}$ *14*

$\boxed{\frac{1}{2}}$ $\boxed{\frac{7}{12}}$ ✓ $\boxed{\frac{5}{8}}$ $\boxed{\frac{2}{3}}$

---

60 Bilal scores **30** out of **50** in a test.
Sadia scores **65%** in the same test.

Who has the higher score?   **Bilal**   (**Sadia**)

✓

W47
5 8 24

Explain how you know.

$$\frac{30 \times 2 = 60}{50 \times 2 = 100} = 60\%.$$

checked

....

....

---

61 Circle the fractions that are equivalent to **0.75**

$\frac{5}{7}$ $\left(\frac{75}{100}\right)$ $\frac{1}{75}$ $\left(\frac{3}{4}\right)$ $\frac{2}{3}$ $\frac{7}{5}$

✓

---

62 Circle the decimal which is equivalent to **one fifth**.

(**0.2**)   **0.3**   **0.4**   **0.5**

✓

Name ....

63 Tom goes to a football match.
He spends $\frac{3}{4}$ of his money on a ticket and $\frac{1}{2}$ of the change on a programme.
He has **£2** left.
How much money did he have to start with?

£ [16]

64 Ayesha makes a cake which weighs **480 g**.
35% of the weight is butter, $\frac{1}{5}$ is flour, **25%** is sugar, $\frac{3}{20}$ is eggs and the rest is cocoa powder.
How many grams of each ingredient did she use?

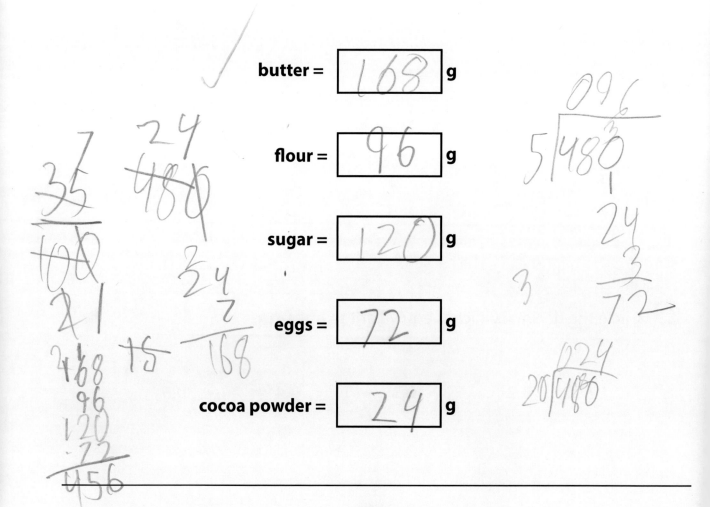

butter = [168] g

flour = [96] g

sugar = [120] g

eggs = [72] g

cocoa powder = [24] g

It may be appropriate for children to use exercise books or paper to record their answers, working out or explanations.

MEASUREMENT

These are all about measurement!

MEASUREMENT

Convert between centimetre and metre

1 The table is **1.2 metres** long. How long is the table in **centimetres**?

120 cm

2 Tom is **172 cm** tall. Bilal is **1.73 metres** tall. Who is taller? Explain your answer. Bilal as 1.73m = 173cm so Bilal is taller

3 The length of a book is **0.32 m**. What is the length of the book in **centimetres**?

32 cm

Checked

4 Lola is **118 centimetres** tall. How tall is this in **metres**?

1.18m

5 Charlie won the **100 metre** race. How far did he run in **centimetres**?

~~1m~~ 10000cm

6 The giraffe is **4.1 metres** tall. How tall is it in **centimetres**?

410 cm

7 Bilal plants an oak tree which is **82 cm** tall. In a year, the oak tree grows another **56 cm**. How tall is the oak tree in **metres** after a year?

1.38m

8 The Clarks' garden is **110 metres** long. The Browns' garden is **65 metres** long. How much longer is the Clarks' garden in **centimetres**?

4500cm

MEASUREMENT

Year 5

Convert between kilometre and metre

1 Bilal runs **3.6 kilometres** in the school fun run. How many **metres** did he run?

3600m

2 The lorry driver travels **9600 metres**. How far has she driven in **kilometres**?

9.6km

3 A farmer's sheep pen is **4.26 km** wide. How wide is the sheep pen in **metres**?

4260   checked

4 The drive to the theme park for the Hughes family is **6200 metres**. How far do they have to drive in **kilometres**?

6.2 km

5 Tom is planning a walk in the hills. The walk is **7.38 kilometres**. How many **metres** is this?

7380 m

6 On holiday in Cornwall, Peter drives **49 km** to the beach. The next day his wife, Maria, drives **48,920 metres** to Land's End. Who drove further? Explain your answer.

Peter as   49km = 49000 m

7 Ayesha jogs **2243 metres** and runs **4359 metres**. How far does she travel altogether, in **kilometres**?

6.602km

2243
4359
6602

8 In the marathon, Lola ran **41.35 km**, Sadia ran **42,195 m**. Who ran further? Explain your answer.

Sadia as   42195 m = 42.195km

Name ....

MEASUREMENT   **Year 5**

Convert between centimetre and millimetre

1 A centipede is **3 centimetres** long. How long is it in **millimetres**?

30mm

2 The length of Bilal's phone is **100 mm**. How long is his phone in **centimetres**?

10 cm

3 A red ant measures **0.5 centimetres**. What is the length of the ant in **millimetres**?

5 mm

4 Mrs Gregson is planting flowers. She puts **12 centimetres** of soil into the plant pot. How many **millimetres** of soil are in the plant pot?

120 mm

5 A worm is **14.2 cm** long. How long is it in **millimetres**?

142 mm

6 A battery is **33 mm** in length. A TV remote needs **2** batteries. What is the total length of the batteries in **centimetres**?

6.6 cm

7 A chocolate bar measures **125 millimetres**. Lola eats **2 cm** of the chocolate bar. How much of the chocolate bar is left? Show your answer in **centimetres**.

10.5 cm

✓ checked

8 Convert **342 millimetres** into **centimetres**.

34.2 cm

25-3-24

Convert between gram and kilogram

1 Mrs Ahmed is cooking dinner. She weighs out **2000 grams** of chicken into a bowl. How much chicken is there in **kilograms**?

2 kg

2 Sadia picks **2.3 kg** of apples from her tree. She converts the weight of the apples into **grams**. What should her answer be?

2300 g

3 Mr Purtill's laptop bag weighs **3.25 kg**. How much does his bag weigh in **grams**?

3250 g

checked

4 Convert **2430 grams** to **kilograms**.

2.43 kg

5 A box contains **7000 grams** of bananas. The zookeeper takes out **4 kilograms** for the monkeys to eat. How many **grams** of bananas are left?

3000

6 Bilal is going on holiday to Australia. His suitcase weighs **18.5 kg**. His sister's suitcase weighs **18,450 g**. Whose suitcase is heavier? Explain your answer.

Bilal as 18.5kg = 18,500g

7 Mr Smythe, the shopkeeper, buys **5 kilograms** of apples, **6.2 kg** of oranges and **4230 grams** of pears from the wholesaler. What is the total weight of the fruit in **grams**?

4350 g 15430 g

8 Sadia is making cakes at the bakery. She has **2300 grams** of flour. She needs **3 kg** of flour altogether. How much more flour does she need in **grams**?

700 g

11.2
4.2.3
6.2
4.23   43.5

10.43

W43
8-7-24
checked

Convert between litre and millilitre

1 A bottle contains **two litres** of fizzy orange. How many **millilitres** of fizzy orange are in the bottle?

2000ml

2 A big bottle of milk holds **4000 millilitres**. How many **litres** of milk are in the bottle?

4L

3 Ayesha fills the sink with **5500 millilitres** of water. How many **litres** of water are in the sink?

5.5L

4 A bucket is filled with **8.2 litres** of water. How much water is in the bucket in **millilitres**?

8200ml

5 An average household uses **153 litres** of water every day. How many **millilitres** of water are used on average?

153000ml

6 Fish Tank **A** holds **4460 millilitres** of water. Fish Tank **B** has **4.8 litres** of water. Which tank holds the larger capacity of water? Explain your answer.

4.46   B

7 Tom drinks **1.5 litres** of water on Monday. His dad drinks **1800 millilitres** of water. Who drinks more water? What is the difference?

Dad by 300 ml or 0.3L

8 A jug of juice is made up of **500 ml** of cordial and **1.5 litres** of water. How many **millilitres** are in the jug in total?

2000ml

Name ....

Convert between units of measure (mixed)

1 A bag of flour weighs **1.2 kg**. What is the weight of the flour in **grams**?

1200 g

2 Bilal walks **2.4 kilometres** to the skate park. How far has he walked in **metres**?

2400 pt

3 Sadia weighs **24.5 kg**. How much does she weigh in **grams**?

24500 g

4 Miss Powell has a **1 litre** bottle of water. She drinks **550 ml**. How much water does she have left? Give your answer in **ml**.

450 ml

5 The size of Tom's phone is **115 mm**. What is the length of Tom's phone in **centimetres**?

11.5 cm

6 Charlie runs **2.2 kilometres** to the shops and then walks another **1300 metres** to the park. How far has he travelled altogether in **kilometres**?

3.5 km

7 Ayesha pours **8.5 litres** of water into the paddling pool. The dog drinks **1500 millilitres**. How much water is left in the paddling pool in **litres**?

7 L

8 Lola's suitcase weighs **23,800 grams**. She is only allowed to take **22 kg** onto the plane. How much over the weight limit is her suitcase? Show your answer in **kg**.

1.8 kg

Convert between units of measure (mixed)

1 Convert **3 litres** of orange juice into **millilitres**.

3000 ml

2 Tom runs **3.5 km** around the park. How far has he run in **metres**?

3500 m

3 The length of an eraser is **50 mm**. What is the length of the eraser in **centimetres**?

5 cm

4 Bilal's school rucksack weighs **2330 grams**. How many **kilograms** does the rucksack weigh?

2.33 kg

5 Sadia is **143 centimetres** tall. How tall is she in **metres**?

1.43 cm

6 Charlie converts **6.92 litres** into **millilitres**. His answer is **692 ml**. Is he correct? Explain your answer. No as 6.92 L = 6920 ml and ml

7 Ayesha is making raspberry jam. She has **2.42 kg** of raspberries. How many **grams** of raspberries does she have?

2 x 420 g

8 The recipe for raspberry jam states that you need **1950 grams** of raspberries. How many **kilograms** of raspberries will Ayesha have left over? Use your answer from Question 7 to help you.

0.487 g

1 7420
7950
0480

Understand and use equivalences between inches and centimetres

1 **One inch** is equivalent to approximately how many **centimetres**?

2.5 cm

2 Write down **three** things that you might measure in **inches**.

jar, cup, book

3 Sadia measures the length of a sweet, which is **2.5 cm**. Approximately, how many **inches** long is the sweet?

4 Lola has **2 inches** of her hair cut off at the hairdresser's. Approximately, how many **centimetres** of her hair have been cut off?

5 cm

5 Ayesha grows **4 inches** in a year. How many **centimetres** has she grown?

10 cm

6 Tom has built a wall with toy bricks that is **10 inches** high. Approximately how tall is the wall in **centimetres**?

25 cm

7 Approximately, how many **inches** are equivalent to **15 cm**?

6

8 **One foot** is equal to approximately how many **centimetres**? (Not the foot at the end of your leg!)

30 60 cm

Understand and use equivalences between pounds and kilograms

1 Approximately, how many **pounds** are equivalent to **1 kilogram**?

2.2

2 Lola needs **2.2 pounds** of sugar to make her dessert for the party. Approximately, how many **kilograms** of sugar does she need?

3 The suitcase weighs **12 kilograms**. Approximately, how many **pounds** does the suitcase weigh?

26.4

4 Approximately, how many **kilograms** are equivalent to **6.6 pounds**?

3kg

5 Tom has **6.5 pounds** of apples. Does he have nearer to **3** or **4 kg** of apples?

3kg

6 Ayesha weighs **33 kilograms**. Sadia weighs **78 pounds**. Who weighs more? Explain your answer.

Sadia as 33kg = 72.6 lbs

7 An enormous cake recipe needs **3 kilograms** of butter and **4 kilograms** of flour. Approximately, how many **pounds** of butter and flour does the recipe need altogether?

15.4

8 A rugby player weighs **90 kilograms**. Approximately, how many **pounds** is this?

198 lbs

33
2.2
660
660
726

w 4 9

Understand and use equivalences between pints and litres

1 Approximately, how many **pints** are in **1 litre**?

1.75 pin

2 Mrs Ahmed buys **2 litres** of milk at the corner shop. Approximately, how many **pints** of milk has she bought?

3.5 pin

3 Ayesha drank **3 litres** of water in a day. How many **pints** did she drink approximately?

5.25 pin

4 Bilal says, "**4 litres** of water are equivalent to approximately **7 pints**." Is he correct? Explain your answer.

yes

5 Sadia uses **4 pints** of water, to water her flowers. Lola waters her flowers with **2 litres**. Who has used the most amount of water for their flowers? Explain your answer.

Sadia as 2l = 3.5 pin

6 Mr Gregson asks his class, "Approximately, how many **litres** are there in **7 pints** of water?" What answer should his class give?

4l

7 The Khan family drink **14 pints** of orange juice in a month. Approximately, how many **litres** of orange juice do they drink in a month?

8l

8 The goldfish pond is filled with **1000 litres** of water. Approximately, how many **pints** of water are in the pond?

1750 pin   ✓ checked

66 22
85   9
194 198

Name ....

Calculate the perimeter of rectangles

1 A rectangle has **four** sides. The length of the rectangle is **6 cm** and the width is **3 cm**. What is the perimeter of the shape?

18 cm

2 The length of a slice of bread is **20 cm**. The perimeter is **70 cm**. What is the width of the slice of bread?

10 cm

3 Bilal has a rectangular hand-held games console which has **four** sides. The length of the console is **14 cm** and the width is **10 cm**. Work out the perimeter of the games console.

48 cm

4 The length of a vegetable patch is **8 metres**. Its width is **4 metres**. What is the perimeter of the vegetable patch?

24 m

5 Each side of a square measures **8 cm**. Tom says, "To work out the perimeter of the square, you can multiply **8** by **4**." Is he correct? Explain your answer.

no yes perimeter = 8+8+9+9 = 24 cm

6 Bilal has forgotten how to work out the perimeter of an oblong. Explain how he should do this.

7 Look at the measurements below:

Rectangle A: Length 8 cm Width 5 cm
Rectangle B: Length 10 cm Width 5 cm

A new shape is made by putting the two rectangles together, end to end. What is the perimeter of the new shape?

46 cm

Calculate the areas of rectangles (including squares)

1 A stamp's length is **3 cm** and the width is **2 cm**. What is the area of the stamp?

$6 cm^2$

2 The sides of a square sweet are **2 cm**. What is the area of the sweet?

$4 cm^2$

3 The length of Ayesha's reading book is **20 cm**. The width of the book is **10 cm**. What is the area of the book?

$200 cm^2 = 2 m^2$

4 The area of a square is **16 m²**. What is the length of each side?

$4 m$

5 Mr Knight's class are learning about area. He asks them to find the length of a rectangle which has an area of **18 cm²** and a width of **3 cm**. What should the answer be?

$6 cm$

6 What is the area of a square with sides measuring **6 cm**?

$36 cm^2$

7 A vegetable patch is **12 metres** long and **8 metres** wide. What is the total area of the vegetable patch?

$96 m^2$

8 A new shape is made by placing a square with **4 cm** sides next to an oblong with a length of **5 cm** and a width of **7 cm**. What is the area of the new shape?

$51 cm^2$

136

Name ....

MEASUREMENT   Year 5

Solve problems converting between the 12 and 24 hour clock

1 Is the time **22:45** in the morning or afternoon? Explain your answer.

afternoon because 00:00 → 12:00 is am and
12-17-3 16 and

2 The time on a **24 hour** clock is **15:30**. What time would this be on a **12 hour** clock?

checked

03:30 pm

3 An analogue clock shows **5 o'clock** in the evening. Write down the time shown on her digital **24 hour** clock.

17:00

4 The class clock reads **13:57**. What would the time be on a **12 hour** digital clock?

01:57 pm

5 Mr Patel goes to work at **8 am**. He finishes work **7 hours** later. What time would his **24 hour** digital clock show when he finishes work?

15:00

6 A football match started at **1.30 pm** and finished at **3.06 pm**. The players arrived back in the changing rooms **5 minutes** later. What time did the **24 hour** clock show?

15:11

7 How many minutes are there from **11.30 am** to **14:45**?

195 mins   195

8 How many minutes are there between **10 to 11** in the morning and **13:20**?

150 00 mins

MEASUREMENT　　　　　　　　　　　　　　　**Year 5**

Solve problems converting between units of time

1 How many **minutes** are in an **hour**?

60

2 It takes Tom **three and a half hours** to complete his homework. How many **minutes** does it take Tom to complete his homework?

210　　　　　　　　　　　　　*checked*

3 Class 5 take part in a **2 minute** sponsored silence. For how many **seconds** do they need to be silent?

120

4 Lola jogs every day for **3 weeks**. How many **days** is this?

21

5 Sadia goes swimming for **one hour** every day for **2 weeks** and **3 days**. How many **hours** does she swim for altogether?

17

6 Class 5 spend **5.5 hours** a day at school. How many **minutes** are they at school for each day?

330

7 There are **24 hours** in a day. How many **minutes** are there in each day?

1440 min

8 Sadia reads for **10 minutes** every night for **one week**. How many **seconds** does she read for in a week?

checked.

4200 sec

Use all four operations to solve problems involving measure

1 If **ten litres** of cola costs **£3.50**, what is the cost of **two litres**?

2 Bilal drinks **1.5 litres** of water. Charlie drinks **350 ml** of water. How much more does Bilal drink than Charlie? Give your answer in **millilitres**.

3 Sadia pours **two** bottles, each containing **1.5 litres** of water, and **eight** bottles, each containing **350 ml** of water, into her fish tank. How many **millilitres** of water does she pour into the tank altogether?

4 A lorry depot orders **30,000 litres** of petrol. If the lorries use up **250 litres** of petrol each day, how long will this supply last?

5 If you cycle **twelve** times around a **1500 m** track, how far do you cycle in total? Give your answer in **kilometres**.

6 How many full laps of a **400 metre** track does Tom need to run, if he is aiming to run for at least **10 km**?

7 What is the total weight of **two** biscuit boxes, each weighing **325 g**, and **two** biscuit boxes, each weighing **0.5 kg**? Give your answer in **kilograms**.

8 Lola buys **3.67 kg** of potatoes. Ayesha buys **10 kg** of potatoes. How many more **grams** of potatoes does Ayesha buy than Lola?

Use all four operations to solve problems involving measure

1 A large jug has a capacity of **6.5 litres**. How many cups of apple juice could be poured from the jug, if each cup contains **250 ml** of apple juice?

2 Each carton of milk contains **1.5 litres**. How many **litres** of milk are there in a box containing **six** cartons?

3 Tom uses **9** oranges to make **one quarter** of a **litre** of orange juice. How many **litres** of orange juice can he make from **72** oranges?

4 Bilal goes on a **5.5 km** run. Charlie runs **2700 metres**. How much further does Bilal run than Charlie? Give your answer in **kilometres**.

5 A daisy is **4 cm** long. What would be the total length of **125** daisies? Give your answer in **metres**.

6 Sadia rides her bike for **0.78 km** and then runs for another **435 metres**. How far does she travel in total? Give your answer in **metres**.

7 Tom buys **half a kilogram** of chicken, **300 grams** of cheese and **200 grams** of butter. What is the total weight of his shopping in **kilograms**?

8 **One** large bag of rice weighs **3.4 kilograms**. What is the total weight of **3** large bags of rice in **grams**?

MASTERING

Measurement

Time to think about measurement!

w 48
10-8-29

1 Fill in the gaps to make these statements correct.

3.6 litres + 700 ml = 5000 ml – [0.7] litres

500 *(200)*

5 m – 300 cm = 40 cm + [~~160~~] cm

25 cm + $\frac{3}{4}$ m = 200 cm – [~~100~~] m

450 g + 1.05 kg = 2 kg – [~~500~~] g

---

2 Little Grove Primary School has **three** fences of different lengths. Tom, Charlie and Bilal measure each fence to compare their lengths. Unfortunately, they each use different units.

Put the fences in order of length, starting with the shortest.

fence 1 **1** 212 m

fence 2 **2** 22100 cm

fence 3 **3** 0.211 km

fence [3] fence [1] fence [2]

shortest **longest**

---

3 Tom has **5** exercise books each **24 cm** long.
He puts them on his table and notices that the **5** exercise books are exactly the same length as **3** ring binders when put end to end next to each other.

How long is **one** ring binder?

books

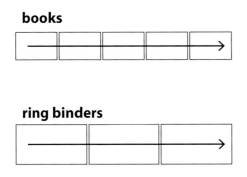

ring binders

cm

4 Lola fits **two** pieces of ribbon together to make part of a border for a picture as shown.
The ribbons overlap by **0.15 m**.
The picture is **96 cm** wide and the shorter piece of ribbon is **0.33 m** long.

How long is the longer piece of ribbon?

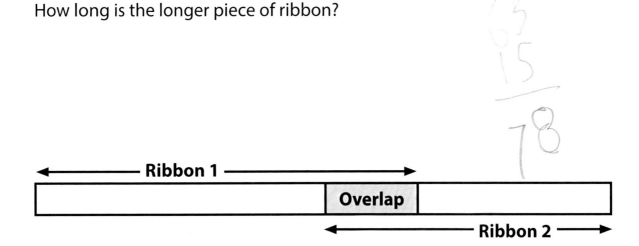

cm

5 I want to put one row of tiles along a wall in my bathroom.
Tiles are either rectangles measuring **5 cm** by **8 cm**, or squares with a side
of **5 cm**.
The length of the wall is **60** cm.

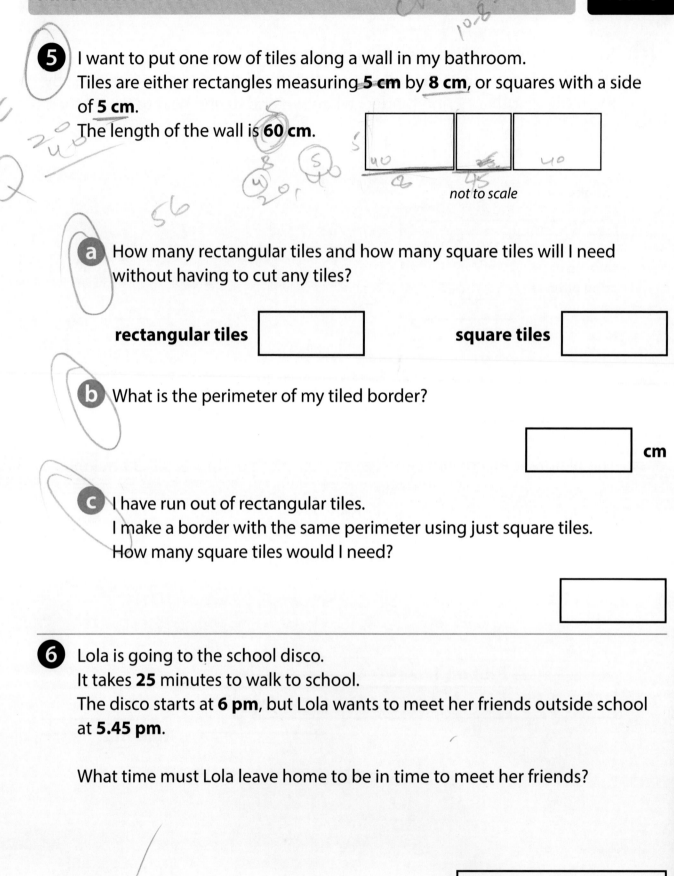

not to scale

a How many rectangular tiles and how many square tiles will I need
without having to cut any tiles?

rectangular tiles [ ]

square tiles [ ]

b What is the perimeter of my tiled border?

[ ] **cm**

c I have run out of rectangular tiles.
I make a border with the same perimeter using just square tiles.
How many square tiles would I need?

[ ]

6 Lola is going to the school disco.
It takes **25** minutes to walk to school.
The disco starts at **6 pm**, but Lola wants to meet her friends outside school
at **5.45 pm**.

What time must Lola leave home to be in time to meet her friends?

5:20 pm

Mr Fletcher wants to buy tablets for Year 5 to use in ICT lessons.
He cannot decide whether to buy full-size tablets or mini tablets.
He knows that the trolley where they are stored cannot hold more than
13 kg.
He does not have scales to weigh the tablets, but he knows that:

• **Two** full-size tablets weigh the same as **three** mini tablets.

• **One** full-size tablet and **one** mini tablet weigh $\frac{3}{4}$ **kg** altogether.

Can Mr Fletcher buy **30** full-size tablets and store them on the trolley?

Explain your answer.

Yes / No

....

....

 Bilal has made a cuboid using **54** small cubes.
The length and the height of the cuboid are the same.

What are the dimensions of all **three** sides?

length = | 3 | cm

width = | 3 | cm

height = | 6 | cm

9 Miss Chen wants to make a small rectangular patio area in her garden.
She has **16** congruent square paving slabs with sides of **1 m**.
When she has laid her patio, she wants to put a border of flowers around
her patio area.

a What is the smallest and largest perimeter she can have using all **16**
paving slabs?

smallest = [16] **m**   largest = [34] **m**

b Miss Chen places the paving slabs in an **8 x 2** arrangement.
She wants to plant roses and pansies as a border.
Roses cost **£3.25** for **4** plants.
Pansies cost **£2.50** for **3** plants.
She needs **9** plants for every metre of border and plants twice as many
roses as pansies.

How much does it cost her?

£ [147.50]

c Miss Chen has **£150** to spend on bedding plants.

How much change should she get?

£ [2.50]

10 Charlie drew a square with sides of **6 cm**.
He said, "I can draw **five** rectangles with the same perimeter as the square,
but the square has the largest area."

Is Charlie correct? | Yes / No |

Explain your answer.

....

....

11 Sadia bought a cube of cheese on Monday.
Each side of the cube was **8 cm** long.

a What is the area of each face of the block of cheese?

64 cm²

b When she got home she wrapped the block of cheese in cling film.
She wrapped the cling film three times around the block so that there
was a triple layer of cling film.

What was the length of the cling film that she used?

96 cm

c On Tuesday, she took off the cling film and cuts a slice, **2 cm** wide,
from one face of the cube. What were the dimensions of the block of
cheese then?

8 cm × 8 cm × 6 cm

d What shape was the block of cheese then?

cuboid

e On Wednesday, she cut off two more **2 cm** wide slices of cheese.
She cut them from the same face as on Tuesday. What were the
dimensions of the cheese that was left?

8 cm × 8 cm × 2 cm

12 Ayesha has a bucket that holds **5 litres**. Lola has a jug that holds **3 litres**.
They want to put **7 litres** of water into a bowl.
How can they measure out exactly **7 litres**?

1,5L + ⅔ 3L

....

....

13 Lola uses regular hexagonal plastic tiles to make a pattern on the floor of
the hall.
The perimeter of **one** tile is **1.5 m**.

She uses black and white tiles to make a design like the one below.

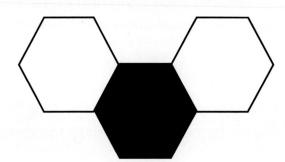

a What is the perimeter of the shape?

3.5 **m**

b She repeats this pattern using **231** tiles altogether. How many black
tiles and how many white tiles has she used?

white = 154   black = 77

c The tiles were wrapped in sealed polythene packs of 20. Once she has
opened the packs that she needs, how many of each colour will she
have left?

white = 6   black = 3

 Mrs Cook has a pile of **25** maths books on her shelf. The pile is **1.25 m** tall. Mr Watts takes **10** maths books to use with his group.

What is the height of the pile of maths books on Mrs Cook's shelf now?

0m75 **cm**

 Sadia has **12** small red cubes.
She puts them together to make a cuboid of sides **2 cm**, **2 cm** and **3 cm**.
Sadia paints the faces of the cuboid blue. When the paint is dry, she takes her cuboid apart again.

When she does this, she notices that some of the small cubes have been painted blue on **two** faces and some on **three** faces.

How many cubes have been painted on:

a **2 faces** 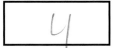 4

b **3 faces**  8

If it helps, you could make the cuboid.

 The ratio of the length : height : depth of a cuboid is **1 : 2 : 3**.
The total surface area is **88 cm²**.

Find the length, height and depth of this cuboid.

You might want to build a cuboid to help you work this out.

length =  14   **height =** 28   **depth =** 42

148   Name ....

17 The diagram shows a large square of side x cm with a smaller square of side y cm cut out of it. If $x = 7$ cm and $y = 3$ cm, what is the perimeter of the new shape?

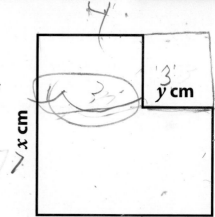

$7 + 7 + 7 + 7$

$7 \times 4 =$

2 ~~48~~   **cm**

---

18 Sadia knows that **1 mile** is approximately equal to **1.6 km**.
She ran **3 miles** every day for a week.
Tom ran **35 km** in the same week.
Who ran the furthest during the week?

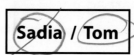

Sadia / Tom

Explain your answer.

....

....

---

19 Mr Watson likes to arrange his exercise books in neat patterns.
He always overlaps each book by **4 cm**.
He has **5** books in a pile.
The total length of all **5** books is **84 cm**.

How long is **one** exercise book?

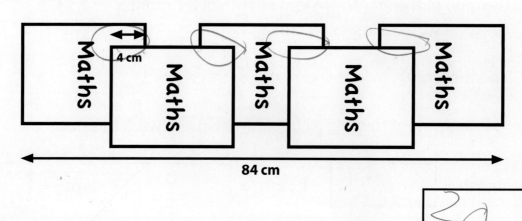

20   **cm**

---

20 The thick line shows the path through a square maze.
Find the length of the path.
Write your answer in metres.

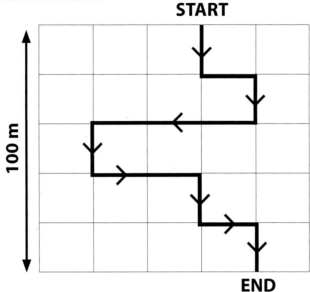

240 m

21 Lola and Tom are each given a bookmark measuring **18 cm** to decorate.
Lola divides her bookmark into **four** equal sections.
Tom divides his bookmark into **3** equal sections.
Lola and Tom put their bookmarks on the table in front of them like this:

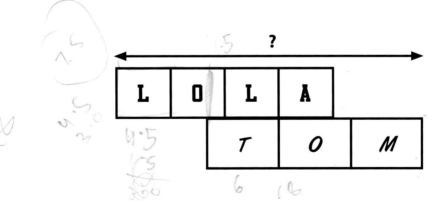

What is the total length of the **two** bookmarks if they are placed like this?

25.5 cm

22 Tom and Bilal ran **1.950 kilometres** altogether.
Tom ran **twice** as far as Bilal.

How far did each boy run in metres?

Tom = 3 900 **m**   **Bilal =** 1950 **m**

23 Charlie had a piece of wire measuring **120 cm**.
He bent the wire to make **4** congruent equilateral triangles.

How long is the side of each triangle?

10 **cm**

24 Bilal made **three** towers using rectangles and octagons.

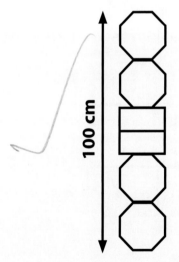 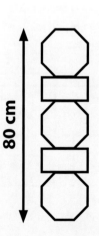 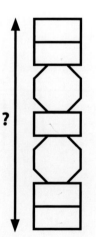

100 cm   80 cm   ?

How tall is the **third** tower that Bilal made?

90 **cm**

25 I have some tiles in the shape of equilateral triangles with a perimeter of **15 cm**.

a If I put **two** triangles together, what is the perimeter now?

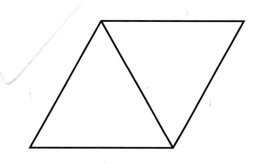

20 | **cm**

b Using two of these triangles, draw a shape that has a perimeter of **30 cm**.

26 A cat has fallen into a well **10 m** deep.
It tries to climb out of the well.
Every **5** minutes, it manages to climb **3 m** up but then slips back **2 m**.

If it keeps climbing at this rate, how long will it take the cat to climb out of the well?

50 | **minutes**

27 Bilal and Charlie are going on holiday.
They are allowed to take **one** bag each weighing up to **23 kg**.
Any bags weighing more than that will have a surcharge of **£10** for every extra **5 kg** or part of **5 kg**.
They weigh their suitcases before they go.
Bilal's suitcase is **three** times as heavy as Charlie's suitcase.
The difference in weight between the two suitcases is **22 kg**.

a How heavy is each suitcase?

Bilal = 33 **kg**   Charlie = 11 **kg**

b How much excess charge will Bilal have to pay?

£ 20

c How could both boys go on holiday without either having to pay a surcharge?

take less stuer or share the stuer between each suitcase

28 A rectangular playground has a **one-metre** high fence around it which needs painting. The playground has an area of **1600 m²**. Paint costs **£4.50** a tin. **One** tin of paint covers **50 m** of fence. Mr Smith pays **£18** for paint and uses it all to paint the fence.

What is the length and width of the fence?

length = 20 **m**   width = 80 **m**

1600 m²

12-8-24 W4 8

29 Charlie draws a square and an equilateral triangle.
The sides of the square are equal in length to the sides of the triangle.
The perimeter of the square is **5 cm**.

What is the perimeter of the triangle?

1·25

4÷5

| 3.75 | cm

---

30 Motorway maintenance are repairing a length of crash barrier. Each
section of crash barrier is **2.5 m** long.
1 km of crash barrier needs to be replaced.

4   10

a How many sections will be needed to replace the crash barrier?

Checked

| 400 |

b Each section costs **£125**. What is the total cost of the new crash
barrier?

£ | 50000 |

---

31 Find the perimeter of this shape.

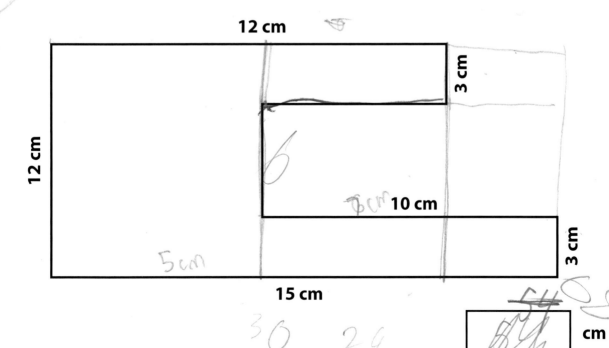

12 cm

3 cm

12 cm

8cm 10 cm

3 cm

5cm

15 cm

30   24

| 54 | cm

GEOMETRY

Properties of shapes / Position and direction

GEOMETRY - Properties of shapes / Position and direction   **Year 5**

Know angles are measured in degrees: estimate and compare acute, obtuse and reflex angles; identify angles

Checked

1 Ayesha has measured an angle and says her answer is **45 degrees Celsius**. Is she correct? Explain your answer.

No as celsius means how hot is it

2 Next, she says that an obtuse angle is bigger than **90°**. Has she explained it fully? Explain your answer.

No as obtuse angle = 90° - 180°

3 What is an angle of **90°** called?

right angle

4 Tom says, "An angle that is bigger than **90°** is acute." Is he correct? Explain your answer.

No as acute angle = 0° ≠ 90°

5 Bilal measures an angle that is acute. It is less than **45°** but more than **43°**. What angle has he measured?

44°

6 What is the name of an angle that is bigger than **180°**?

reflex

7 Correct the following: Angles at a point measure ~~300°~~ in total.

360°

8 Complete the sentence below:

Miss Heathcote made half a turn which is ...*180*... °

Name ....

W 37   25-5-24

Know angles are measured in degrees: estimate and compare acute, obtuse and reflex angles; identify angles

1 Lola is going to measure an angle. She picks up a ruler. What mathematical instrument should she use to measure angles? *Checked*

protractor

2 Is an angle of **70°** acute, obtuse or reflex?

3 An angle that is more than **90°** but less than **180°** is called reflex. Is this statement correct? Explain your answer.

No it is called obtuse

4 Sadia measures an angle of **90°**. What is the name of the angle she has measured?

Acute right angle

5 What is the name of an angle which is more than **180°** but less than **360°**?

reflex

6 Sadia makes **one** full turn. How many **degrees** does she turn?

360°

7 Tom measures **three** angles at a point on a straight line. What should the total number of **degrees** be?

180°

8 Lola measures the following angles in a shape:

obtuse   acute   acute

Angle 1 - **100°**   Angle 2 - **35°**   Angle 3 - **45°**

For each angle, write whether it is acute, obtuse or reflex. What shape has Lola measured?

triangle

GEOMETRY - Properties of shapes / Position and direction   Year 5

Use the properties of rectangles to deduce related facts and find missing lengths and angles

1 Ayesha knows that a rectangle is a parallelogram. What does this tell you about the opposite sides of a rectangle?

they are the same/parralel

2 A rectangle is a **four-sided 2D** shape. How many right angles does a rectangle have?

4

3 Tom draws an oblong with an area of **18 cm²**. The width is **3 cm**. Can you work out the length of the rectangle?

6 cm

4 Sadia has to measure **one** angle of a rectangle. What should her answer be? Explain your answer. *90° because a rectangle has the interior angles os 360° degrees 360° ÷ ... 90° = 90° for each angle*

5 Bilal works out that the area of a rectangular carpet is **15 m²**. Can you work out the width and length of the carpet?

L = 5 m W = 3 m

6 The area of a rectangle is **70 cm²**. The width measures **7 cm**. What would the length of the rectangle be?

10 cm²

7 **One** side of a rectangle measures **8 cm** and the perimeter is **30 cm**. Give the lengths of the **3** other sides.

8 cm, 7 cm, 7 cm

8 Mr Mahnoor tells his class the perimeter of a rectangle is **34 cm**. The length is **12 cm**. What is the width of the rectangle?

5 cm

w38   2-6-24
checked

Distinguish between regular and irregular polygons based on reasoning about equal sides and angles

1 A regular polygon is a **2-dimensional** shape. Describe a regular polygon, using an example.

Polygon 2D closed sides
angles
Regular

2 Is a square a regular or irregular polygon? Explain your answer. *regular as there are four equal sides and four equal angles*

3 Sadia measures the angles of a triangle. Each interior angle is **60°**. Is the triangle a regular or irregular polygon?

(regular)

4 Is an oblong a regular or irregular polygon?

(irregular)

5 Tom states, "Scalene and isosceles triangles are regular polygons." Is he correct? Explain your answer. *No as they dont have equal sides*

6 A hexagon's perimeter is **36 cm**. **One** side measures **6 cm**. Could the hexagon be regular? Explain your answer. *yes as*

6 X 6 = 36 cm

7 **One** side of an octagon measures **4 cm**. The perimeter is **30 cm**. Could the octagon be regular? Explain your answer. *No as 30 is not a multiple of 4*

W38   2-6-24

Understand the language of reflection and translation

checked

1 Ayesha uses a mirror in her maths lesson. Is she learning about (reflection) or translation?

2 Charlie wanted to move a shape up and along a coordinate grid. Would he reflect or translate it? Explain your answer. *traslate as he is moving it*

3 Lola uses a mirror to draw a new shape. Has she translated or (reflected) the shape?

4 Bilal says that to translate a shape you are simply moving the shape. Is he correct? Explain your answer. *yes*

5 Miss Dotty draws a shape on a coordinate grid. She asks her class to move the shape **three** squares up and **two** squares left. Will her class reflect or (translate) the shape?

6 Charlie says, "When I reflect a hexagon, the reflected shape is different." Can Charlie be correct? Explain your answer. *No sass it is reflected hexagon*

7 Describe the difference between the terms **reflection** and **translation**. Use examples to illustrate your explanation.

Reflection: *translation*

GEOMETRY - Properties of shapes / Position and direction   **Year 5**

Describe the features of shapes

1 What is the name of a **3D** shape that has a square base and **four** triangular sides?

square based pyramid

2 Draw a **2D** quadrilateral which has no right angles.

3 How would you describe an isosceles triangle to somebody who had never seen one? A triangle with two equal sides and angles and one different side and angle.

4 Tom drew a sketch to show the net of a cube. Use your book or paper to draw what you think his sketch looks like.

5 Lola imagines a cuboid and counts its faces. How many faces does she count?

6

6 Describe a cube using the words faces, edges and vertices.

6 faces, 12 edges and 8 vertices

7 Write down the name of **one 3D** shape that has **5** faces.

square based pyramid   Checked

8 Charlie was asked by his teacher to write a description of a pentagonal prism. What should he have written?

12 faces, 15 edges and 10 vertices

W40   15-6-24

checked

Describe the features of shapes

1 Use a ruler to draw a pair of parallel lines.

2 Use your ruler to draw a pair of lines that are perpendicular to each other.

3 Lola counted the pairs of parallel lines that are in a square. How many did she find?

2 pairs

4 A quadrilateral has **one** pair of parallel lines and **one** line of symmetry. What is the name of the quadrilateral? rhombus trapezium

5 Ayesha described a scalene triangle to her friend. What did she say?

all the sides are different length

6 How do you think the dictionary would describe an equilateral triangle?

all the sides are the same. each angle = 60°.

7 Bilal drew the net of a square-based pyramid. How many triangles does his net have?

4

8 Tom named a **3D** shape that had exactly **12** edges. What shape do you think he might have named?

cube

GEOMETRY - Properties of shapes / Position and direction   **Year 5**

Describe the features of shapes   *checked*

1 Sadia tries to list **4** different shapes which are quadrilaterals. What shapes should she have on her list?

square rectange rhombus   Parreldogran

2 Write down **2** properties of a square.

all same sides   4 right angles

(3) Is a square a rectangle? Explain your answer, by explaining the properties of squares and rectangles. *No as in a rectangle there is two different length of side*

4 Bilal says that a quadrilateral can have **three** right angles and **one** angle that isn't a right angle. Is he correct? Explain your answer. *No*

5 Ayesha drew the net of a cuboid. Use your book or paper to draw what you think her net looked like.

6 How many faces, edges and vertices are there on a tetrahedron?

F = 4   E = 6   V = 4

7 Write down how a dictionary might describe a triangular prism. *3D, 5 sides, 9 edges, 6 vertcies*

8 Investigate the difference between the definitions of a trapezium in the US and the UK.

It may be appropriate for children to use exercise books or paper to record their answers, working out or explanations.

MASTERING

Geometry

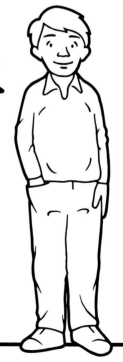

These are tough so ask for help if you need it!

W42
27-24

1 **a** Charlie faces north.
He turns through **135°** clockwise.
In which direction is he now facing?

SE

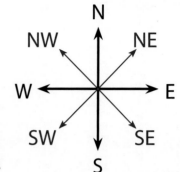

b Charlie faces north and wants to face south-west.
How many degrees and in which direction must
he turn?

135° anti-clockwise

c Charlie faces north-west.
How many degrees anti-clockwise must he turn in order to face south?

135 °

2 Look at nets **A** and **B**.
Tick (✓)the one that will fold to make a cube and cross (✗) that will fold to
make a cuboid.

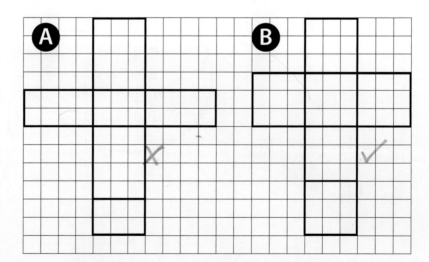

Explain how you know.

If all the sides are
the same

Name ....

3 Sadia has drawn a shape. She says, "I have drawn a rectangle."
Lola says, "You have drawn a square."

Explain how both girls might be correct.

....*They are both quadrilaterals*..

....

4 Here is one angle of an isosceles triangle.
What could the other angles be?
Can you find **two** solutions?

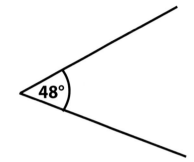

48°

Solution 1 = | 66 |° and | 66 |°   122

Solution 2 = | 984 |° and | 48 |°   135

5 Tick the shape that has just **one** pair of parallel lines.

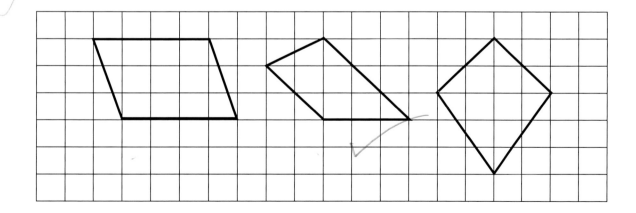

6 Bilal says, "I know this hexagon is regular because it has six lines of symmetry."

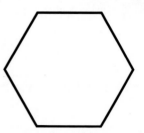

a Is he correct?   Yes / No

b What else do you know about the properties of regular hexagons?

IDK be equal sides 6 equal angles

7 Ayesha drew a rectangle with an area of **12 cm²** on a grid. She translated the rectangle **2** squares right and **2** squares down. Two of the coordinates of the translated rectangle are **(4,4)** and **(7,4).** What could the coordinates of the original rectangle be?
Can you find the coordinates for **two** possible solutions?

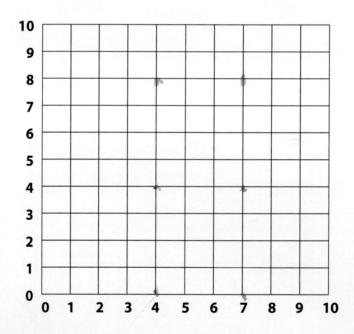

Solution 1 =

(2 , 64)   (5 , 10)

(5 , 6)   (2 , 10)

Solution 2 =

(2 , 2)   (5 , 6)

(5 , 2)   (2 , 6)

Name ....

8 This is a net of a cube. It is missing **one** face.
Complete the net by adding the **6th** face.

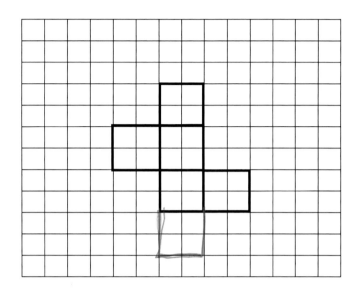

---

9 Sadia drew three congruent rectangles on a grid. The sides of the rectangles are parallel to the axes.

Label the vertices of rectangle **B** with their coordinates.

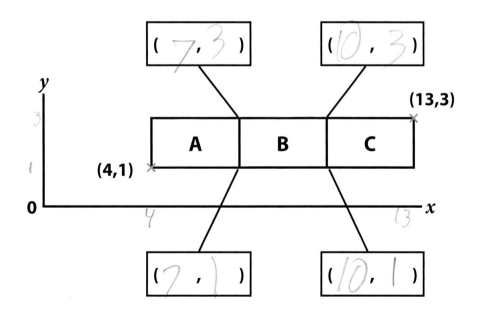

(7 , 3)   (10 , 3)

(13,3)

A   B   C

(4,1)

(7 ,)   (10 , 1)

10 Look at these shapes.

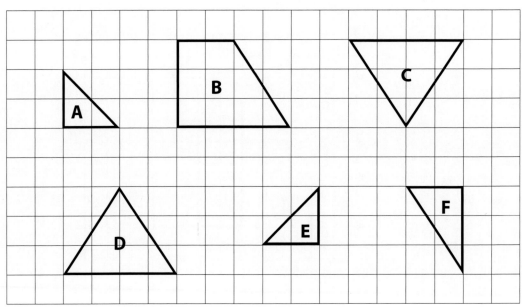

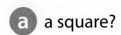

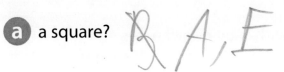

Which **two** shapes fit together to make:

a a square?   ~~B~~ A, E

b a rectangle?   B, F

c a rhombus?   D, C

11 Join the dots to draw as many triangles as possible on the squares below.
Label each triangle as equilateral **(E)**, isosceles **(I)** or scalene **(S)**.
Mark any equal sides and right angles.

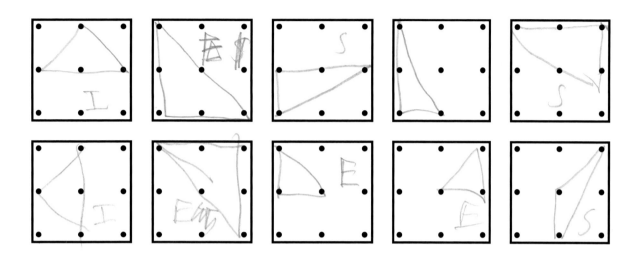

12 I have drawn a right-angled scalene triangle. Work out the size of each
unknown angle.

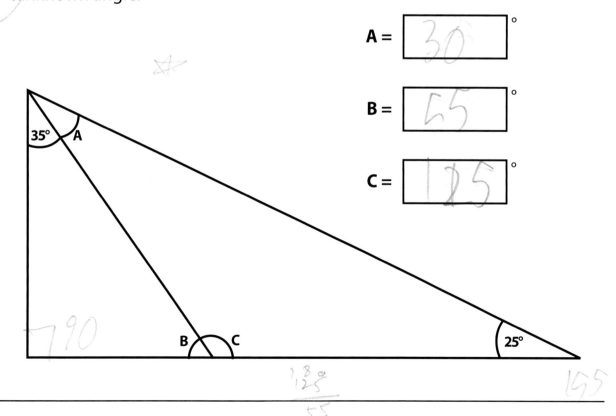

A = 30 °

B = 55 °

C = 125 °

13 Charlie is building a shed in his garden.
The roof supports are made of lengths of wood like this.

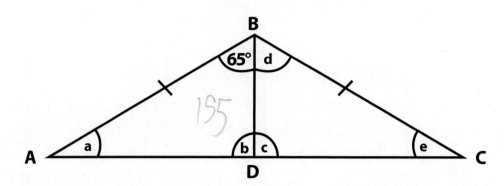

a What type of triangle has Charlie made?

isosceles

b How many right-angled triangles are there?
Mark the right angles on the diagram.

2

c Which side is the same length as **AB**?

B,C

d Which line is perpendicular to **BD**?

A,BC

e Label all the missing angles.

a = 25°   b = 40°   c = 40°

d = 65°   e = 25°

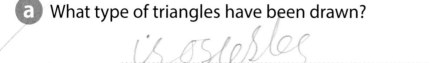
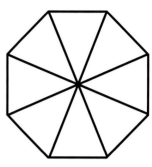

14 A regular octagon is divided into **8** congruent triangles like this.

a What type of triangles have been drawn?

...is osceles...

b What is the size of **one** of the angles at the centre?

45 °

c A regular polygon is divided into congruent triangles in the same way. One angle at the centre is **30°**. How many sides does the polygon have?

8/300

12

---

15 **Two** angles of different triangles are given below. Decide whether each triangle is equilateral **(E)**, isosceles **(I)** or scalene **(S)** or right-angled **(R)**. Explain your answers.

a 60°, 60°   E   because ....

b 55°, 70°   I   because ....

c 115°, 35°   S   because ....

d 25°, 65°   R   because ....

---

16 The diagram shows an isosceles triangle. Find the missing angles.

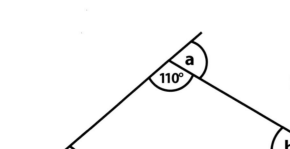

a = 70 °   c = 1545 °

110°

b = 35 °   d = 35 °

e = 145 °

STATISTICS

These are all about statistics!

Solve comparison, sum and difference problems using information presented in a line graph

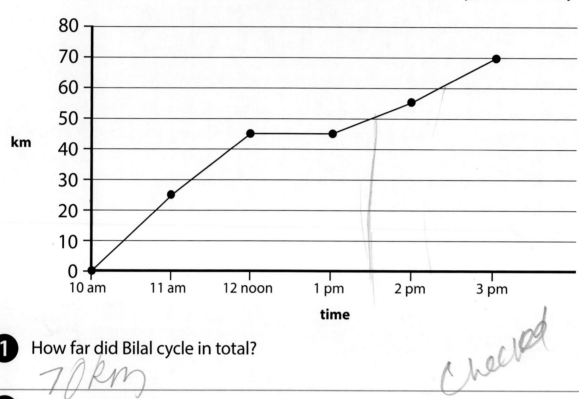

A graph to show the number of kilometres Bilal cycled in a day

1 How far did Bilal cycle in total?

70km

checked

2 How far did Bilal cycle during the first **hour**?

25 km

3 What do you think happened between **12 pm** and **1 pm**?

he took a break

4 By approximately what time had Bilal cycled **50 km**?

1:30 pm

5 Did Bilal travel further between **11 am** and **12 noon** or between **2 pm** and **3 pm**?

11am-12 noon NOON

6 Approximately, how long did it take Bilal to cycle the last **ten kilometres** of his journey?

45 mins

Solve comparison, sum and difference problems using information presented in a line graph

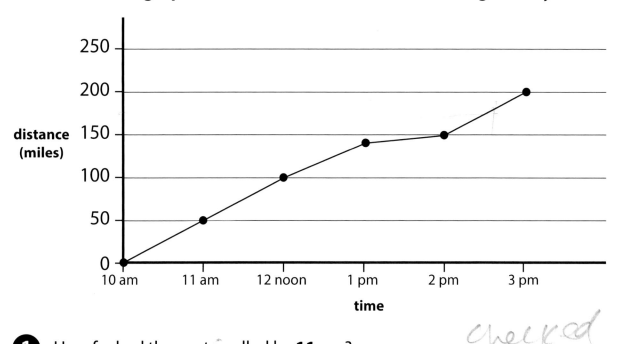

A graph to show how far a car travels during the day

1 How far had the car travelled by **11 am**?

checked

50 miles

2 How long did it take to travel **100 miles**?

2h

3 At approximately what time had the car travelled **175 miles**?

2:30pm

4 How far had the car travelled by **2 pm**?

150 miles

5 How long did it take to travel the last **50 miles**?

1h

6 Between which hours was the least distance covered? Give a possible explanation for this.

1pm - 2pm - traffic lunch

172

Name ....

W-36
27-5-24

Complete, read and interpret information in tables, including timetables

Look at this train timetable.

Checked

	Train A	Train B	Train C
Birmingham	10:45	11:10	12:15
Coventry	10:55	11:20	-
Leamington Spa	11:25	-	12:55
Banbury	11:45	12:00	13:20
Oxford	12:05	12:15	13:40
Reading	12:30	-	14:05
Slough	12:50	12:55	14:25

1 Ayesha arrives at Birmingham Station at **10:30**. How long does she have to wait for the next train? 15m

2 Tom gets on the **11:20 (Train B)** at Coventry. What time does he arrive in Oxford? 12:15

3 How long does the **12:55 (Train C)** from Leamington Spa take to reach Reading? 1h 10m

4 Charlie arrives at Oxford Station at **12:55**. How long does he have to wait for the next train to Slough? 1h 10m 45mins

5 How many stops does **the 11:10 (Train B)** from Birmingham make before reaching Slough? 3

6 Lola arrives in Leamington Spa at **11:30** and waits for the next train to Reading. What time does she get to Reading? 14:05

7 Which train is the fastest from Birmingham to Slough? B

Name ....

Complete, read and interpret information in tables, including timetables

Look at this train timetable.

	Train A	Train B	Train C	Train D
Tiger Street	09:20	11:35	13:45	15:50
Eagle Close	09:25	11:40	13:50	15:55
Jones Station	09:35	11:50	-	16:05
Black Road	09:45	-	14:00	16:15
Ron Avenue	09:50	11:55	-	16:20
Howard Close	09:55	12:00	14:05	16:25
Socton	10:05	12:05	14:10	16:35

Checked

1 Which train is quickest in going from Tiger Street to Socton? C

2 How many times does the **11:40 (Train B)** from Eagle Close stop before arriving at Ron Avenue?

3 What time does the **09:35 (Train A)** from Jones Station arrive at Howard Close? 9:55

4 How long does the **15:55 (Train D)** from Eagle Close take to reach Black Road? 20 min

5 If you want to be at Howard Close by **12:30**, which train should you catch at Jones Station? B

6 If you arrive at Jones Station at **3.40 pm**, how long do you have to wait for the next train? 25 min

7 If you arrive by train at Howard Close at **16:25**, how long did your journey take from Eagle Close? 30 min

Complete, read and interpret information in tables, including timetables

Look at this bus timetable.

	Bus A	Bus B	Bus C	Bus D
High Road	08:40	09:30	10:45	12:15
Angel Close	08:50	09:40	10:55	12:25
Grump Lane	09:05	-	11:10	12:30
Chair Street	09:10	09:55	11:15	-
Fran Close	09:20	10:05	11:25	12:40
Bryant Lane	09:35	-	11:40	12:55
Lew Street	09:40	10:20	11:45	13:00

1 How long does the **08:40 (Bus A)** from High Road take to reach Lew Street?

2 Which bus is quickest at getting from High Road to Lew Street?

3 Sadia catches the **12:25 (Bus D)** from Angel Close. How many times does she stop before she reaches Bryant Lane?

4 You are **ten minutes** late for the **08:40 (Bus A)** at High Road. You catch the next bus. What time do you arrive at Fran Close?

5 How long does the **9:55 (Bus B)** from Chair Street take to reach Lew Street?

6 You arrive at Chair Street bus stop at **11 o'clock**. How long do you have to wait for the next bus?

7 Bilal walked to the Grump Lane bus stop and arrived there at **08:55**. How many minutes is there between the time he got to Grump Lane and the time he arrived at Lew Street?

Complete, read and interpret information in tables, including timetables

Look at this school timetable for Year 5.

	Monday	Tuesday	Wednesday
English	9:00	9:00	9:30
Break	10:30	10:30	11:00
Maths	10:45	10:45	11:15
Lunch	12:00	12:00	12:15
History	13:00	13:00	-
Geography	14:15	14:00	-
Science	-		13:15

1 Which **two** lessons do Year 5 have every Monday, Tuesday and Wednesday?

_____

2 What time does Maths start on Tuesday?

_____

3 How long is the English lesson on Wednesday?

_____

4 On which day do Year 5 study Science?

_____

5 How long is lunchtime on Monday and Tuesday?

_____

6 Sadia is **10 minutes** late to Maths on Tuesday. How long was she in the lesson for in total?

_____

7 School finishes at **15:15**. How many **hours** are Year 5 in school for on a Monday? (They stay in school for lunch.)

_____

Decide which representations of data are most appropriate and why

1 Bilal carried out a survey of his friends to find out their favourite pizza. What would be the best way to represent his data? Explain why this is the best choice.

2 Mr Carter asked Class 5 to record the temperature in class every **30 minutes** during **one** day. Which sort of graph should they choose to record the information?

3 Sadia records how many hours of TV she watches each day, for a week. She wants to record her data in a pictogram. Is this the best form to represent her data accurately? Explain your answer.

4 Year 5 are recording the amount of rainfall within a month. They record their data each day. What would be the best graph to represent the amount of rainfall? Why?

5 Would a line graph be the best graph to show the data below? Explain your answer.

Favourite type of film in Class 5

Comedy	Romance	Scary	Adventure	Sci-fi
8	3	1	10	2

MASTERING

Statistics

Turn your brain up to full power and you will solve these!

MASTERING - Statistics

Year 5

1 This is a train timetable showing the times of trains between London Euston and Birmingham New Street.

Trains

	1	2	3	4	5	6
London Euston	0823	0843	0903	0923	0943	1003
Watford Junction	0837			0937		
Milton Keynes Central		0913		0953		
Rugby			0951			1051
Coventry	0922	0942	1002	1022	1042	1102
Birmingham International	0933	0953	1013	1033	1053	1113
Birmingham New Street	0945	1008	1027	1045	1108	1127

a How many trains stop at Milton Keynes Central between **0823** and **1003**?

2

b Why do you think there are gaps on the timetable?

There are places where the trains dont stop

c How many trains leave London Euston between **0815** and **1015** altogether?

6

2 Charlie needs to be in Milton Keynes Central at **0915**.
Which train should he catch from London Euston?

✓ checked

T 2

Name ....

Trains

	1	2	3	4	5	6
London Euston	0823	0843	0903	0923	0943	1003
Watford Junction	0837			0937		
Milton Keynes Central		0913		0953		
Rugby			0951			1051
Coventry	0922	0942	1002	1022	1042	1102
Birmingham International	0933	0953	1013	1033	1053	1113
Birmingham New Street	0945	1008	1027	1045	1108	1127

3 Lola arrived at Birmingham International at **1033**. What time did she leave London Euston?

> 0923

4 **a** How long does the **0843** train take to travel from London Euston to Birmingham New Street?

> 1hr 25min

b Which other train takes the same time to travel from London Euston to Birmingham New Street?

> train 5

c Which two trains take **3 minutes** less than the **0843** to travel from London Euston to Birmingham New Street?

> train 1 **and** train 4

d Why do you think some trains take longer than others to travel between London Euston and Birmingham New Street?

> some have less stops

Trains

	1	2	3	4	5	6
London Euston	0823	0843	0903	0923	0943	1003
Watford Junction	0837			0937		
Milton Keynes Central		0913		0953		
Rugby			0951			1051
Coventry	0922	0942	1002	1022	1042	1102
Birmingham International	0933	0953	1013	1033	1053	1113
Birmingham New Street	0945	1008	1027	1045	1108	1127

5 Tom's dad wants to travel from Watford Junction to Birmingham New Street for a meeting that starts at **1015**.

a What is the latest train that Tom's dad can catch from Watford Junction?

 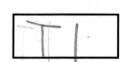

0837

b How many minutes will Tom's dad have to wait before his meeting starts?

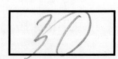 30 **mins**

6 Sadia catches the **0903** train from London Euston.
It leaves **3** minutes late and is delayed for **15** minutes more at Rugby.
What time does the **0903** train actually arrive in Coventry?

Checked

 1020

This bar chart shows the number of texts and emails Mr Wright sent each day last week.

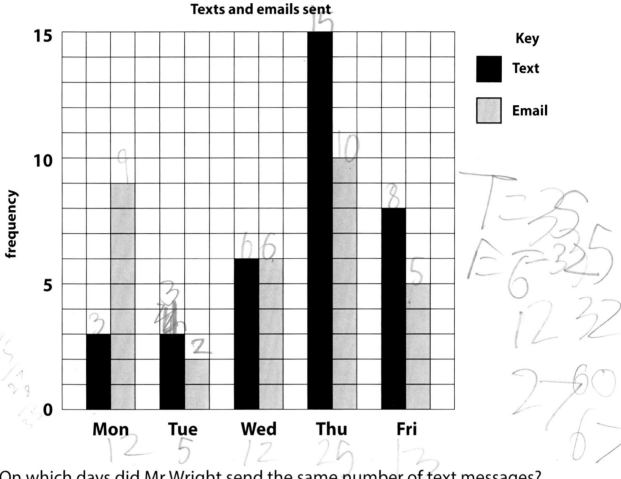

7 On which days did Mr Wright send the same number of text messages?

| Mon | **and** | Tue |

8 On which day did Mr Wright send **three** times as many emails as text messages?

| Mon |

9 On which day did Mr Wright send the same number of text messages and emails?

| Wed |

N44

10 How many more text messages did Mr Wright send than emails?

checked

3

11 On which **two** days did Mr Wright send the same total number of text messages and emails?

Mon **and** Wed

12 On one day, Mr Wright organised the school disco. Which day do you think this was?

Explain your answer.

Thurs as he sent the most messages that day

13 Mr Wright has a monthly contract which allows **150** text messages a month?

Would Mr Wright be able to send the same number of text messages each week as he has this week and stay within his limit?

Yes / No

Explain your answer.

he has sent 35 texts this week. 35×4=140

14 **a** On which day was there the biggest difference between text messages and emails sent?

Mon

b Did Mr Wright send more text messages or emails that day?

Email

c How many more?

6

15 How many text messages on average did Mr Wright send each day?

7

16 During the next week, Mr Wright loses his phone for **three** days.
He sends only $\frac{1}{5}$ the number of text messages he sent this week, but **twice** as many emails.
How many text messages and emails did Mr Wright send during the next week?

text messages = 7

emails = 64

W 44
15·7·24
Checked

17 On Monday, Sadia was not feeling well.
Her mother took her temperature and made her stay in bed until she was feeling better.
This line graph shows Sadia's temperature during the day.

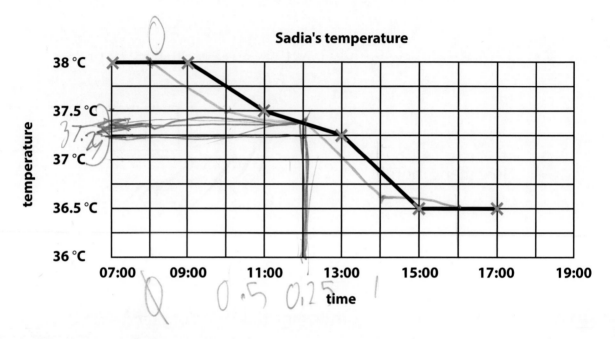

Normal temperature 36.5 °C

Raised temperature   37.5 °C

a How often did Sadia's mum take her temperature?

every 2hrs

b For how many hours was Sadia's temperature raised?

8/6 **hours**

c What happened to Sadia's temperature between **07:00** and **09:00**?

it stayed the same

d Between which times was the biggest fall in Sadia's temperature?

3:00   **and**   15:00

e Why do you think there is no reading for Sadia's temperature at **19:00**?

her temperature is normal

f At what time did Sadia's temperature reading first register normal?

15:00

g What was the difference between Sadia's highest and lowest temperature?

1.5 °C

h What do you think Sadia's temperature was at **12:00**?

37.3 °C

i Sadia's dad also took her temperature during the day. Here are his readings:

08:00	38 °C
10:00	37.5 °C
12:00	37.3 °C
14:00	36.7 °C
16:00	36.5 °C

checked

Plot these readings on the graph.

j How could you change the graph to make it easier to plot these readings?

make the times every hour

18 The pie charts show the different drinks children in Year 5 and 6 chose on the school trip.
There are **two** classes in Year 5 and **one** class in Year 6.
There are **24** children in each class.
The children could choose orange, blackcurrant or water.

Drinks chosen by Year 5   **Drinks chosen by Year 6**

a Water was the most popular drink in both years. Label the pie charts to show this.

b Lola says, "The pie charts show that the same number of children in Year 5 and Year 6 chose water." Is Lola correct? Yes / No

Explain your answer.

Year 5 has more classes

c **Twice** as many children chose blackcurrant than orange in Year 5. Label the sections showing orange and blackcurrant on both pie charts.

d How many more children chose blackcurrant in Year 5 than in Year 6?

Checked

10

e Tom says, "The pie charts show that fewer children in Year 5 chose orange than in Year 6." Is Tom correct? **Yes / No**

Explain your answer.

There are more children in year 5

f There are **60** children in Year 4.
10 children chose orange juice, **20** children chose water and **30** children chose blackcurrant juice.
Complete the pie chart to show this information.

Remember to label each sector accurately and give your pie chart a title.

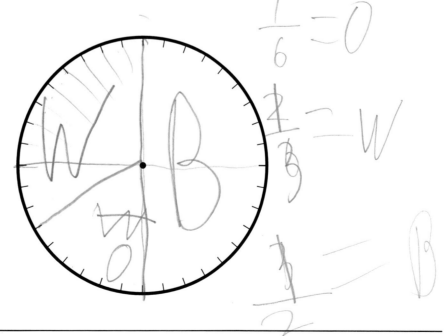

$\frac{1}{6} = 0$

$\frac{2}{6} = W$

$\frac{1}{2} = B$

INVESTIGATION

Sports Day

Year 5 are helping with preparations for the KS2 sports day.

1 There are **four** classes in KS2. There are **twenty four** children in each class.
How many children are there in KS2?

2 The children are divided into **three** houses with an equal number of
children in each house: Livingstone, Nightingale and Scott.
How many children are there in each house?

3 The Year 3 and 4 children sit on large mats to watch the races.
Four children have room to sit in **a metre** square.
The mats measure **2 m** by **2 m**.

How many mats are needed for Year 3 and 4 to sit on?

4 Year 5 and 6 sit on benches.
Each bench measures **2 m**. There is a space between each bench.

Each child needs **30 cm** to sit comfortably on the bench.

How many benches are needed for Years 5 and 6?

5 Use these clues to work out how many Year 5 boys and girls there are in each house.

- There are **8** children in each house.

- There are **2** more boys than girls in Year 5.

- **Three quarters** of the children in Livingstone are boys.

- The numbers of boys and girls in Nightingale are both prime numbers.

- There are **three** times as many girls in Scott than boys.

	Livingstone	Nightingale	Scott	Total
boys				
girls				
total				

Mrs Cook goes from school to the supermarket to buy some orange, blackcurrant and lemon juice, then to the shop to pick up the medals and trophies for the children. This travel graph shows her journey.

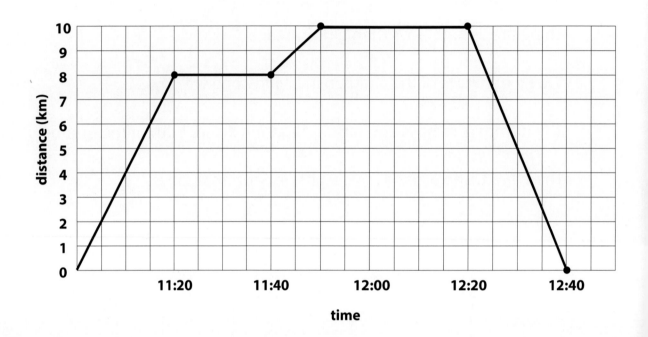

6 How far is it from the school to the supermarket?

⬜ **km**

7 How long did Mrs Cook spend in the supermarket?

⬜ **minutes**

8 How long did it take to travel to the trophy shop from the supermarket?

⬜ **minutes**

9 How much longer did Mrs Cook spend in the trophy shop than in the supermarket?

⬜ **minutes**

10 The events begin at **12:30**.
Mrs Cook allows the following times for each event.
Each event starts straight after the previous one with no gap.
No event can start until the previous one has finished.

	Long jump	Ball throw	Sprint
Year 3	15 minutes	12 minutes	5 minutes
Year 4	20 minutes	15 minutes	7 minutes
Year 5	25 minutes	15 minutes	9 minutes
Year 6	25 minutes	17 minutes	9 minutes

How long should the Sports Day last?

11 Here is the timetable for the events.
The long jump is the **first** event and the sprint is the **last** event.

Complete the timings so that Mrs Cook can keep track of the events.

	Year 3	Year 4	Year 5	Year 6
long jump	12:30	12:45		
ball throw				
sprint				
			FINISH	15:24

The children in Year 5 compete to gain points towards a Triathlon award.

They have to complete **one** jump event, **one** throw event and **one** run event.

The events the children are going to compete in are the long jump, ball throw and **75 m** sprint.

The children gain points as shown in this table:

		Long jump	Ball throw	75 m sprint
BOYS	5 points	5.10 m	27.00 m	12.5 s
	4 points	4.60 m	23.00 m	13.5 s
	3 points	4.00 m	17.00 m	15.0 s
	2 points	2.40 m	11.00 m	17.0 s
	1 point	1.00 m	1.00 m	21.0 s
GIRLS	5 points	4.80 m	18.00 m	12.8 s
	4 points	4.40 m	15.00 m	13.8 s
	3 points	3.60 m	11.00 m	15.3 s
	2 points	2.40 m	7.00 m	17.3 s
	1 point	1.00 m	1.00 m	21.0 s

The boys in Year 5 start the Sports Day with the long jump. Here are their distances.

12 Order their distances from shortest to longest.

Child	Distance
Tom	4.65 m
Ben	3.97 m
Callum	2.39 m
Daniel	3.48 m
Edward	2.27 m
Charlie	4.06 m
George	5.11 m
Hudayfa	4.02 m
Issaa	3.17 m
Jaiden	4.56 m
Karamveer	3.79 m
Luke	1.37 m
Bilal	3.71 m

Child	Distance
Luke	1.37 m

13 Tom's mum is looking at the programme of events. Tom has designed a logo for the cover.

How many triangles can Tom's mum see in the design?

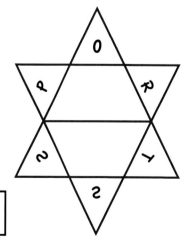

Mrs Cook keeps a record of the points each child receives with the name of the house that they are in.

14 Complete the points for the boys' long jump.
(The rest will be completed later.)

	Child	House	Long jump	Ball throw	75m sprint	Total points	Triathlon award
BOYS	Tom	L	4 points				
	Ben	N					
	Callum	N					
	Daniel	L					
	Edward	L					
	Charlie	S					
	George	L					
	Hudayfa	N					
	Issaa	L					
	Jaiden	L					
	Karamveer	S					
	Luke	N					
	Bilal	N					
GIRLS	Naynaa	N					
	Lola	S					
	Ayesha	S					
	Rachel	S					
	Ebony	L					
	Teghan	L					
	Ursula	N					
	Sadia	S					
	Wanita	N					
	Yvonne	S					
	Zahra	S					

⓯ The girls complete the long jump. Here are their results:

Child	Distance	Rounded
Naynaa	2.55 m	
Lola	4.81 m	
Ayesha	2.49 m	
Rachel	4.38 m	
Ebony	3.45 m	
Teghan	2.05 m	
Ursula	4.02 m	
Sadia	4.39 m	
Wanita	4.78 m	
Yvonne	3.25 m	
Zahra	3.58 m	

⓰ Round their distances to the nearest **metre** to complete the table above.

⓱ Enter the points for each of the girls into the table on page 194.

18 Mrs Cook rolls **two** dice to decide whether the girls or the boys are going to go **first** in the ball throw. The boys will go first if the total of the **two** numbers is even. The girls will go **first** if the total of the **two** numbers is odd.

Who is most likely to go first in Year 5?

Explain your answer.

....

....

....

The total on the dice is **11**, so the girls go first. The dot shows where the ball landed.

Their throws are shown here:

19 What fraction of the girls threw over **15 m**?

20 What fraction of the girls threw less than **10 m**?

21 What type of triangle is used to draw out the shape needed for the ball throw?

Explain how you know this.

....

....

22 Use the diagram to decide which throw matches each measurement in this table.

Measurement	Girl
6.19 m	
7.46 m	
10.06 m	
12.16 m	
13.29 m	
13.84 m	
14.92 m	
16.24 m	
16.43 m	
18.03 m	
19.76 m	

23 Here are the boys' measurements. Fill in the points for the girls' and boys' ball throws into the table on page 194.

Boy	Measurement
Tom	25.46 m
Ben	17.97 m
Callum	27.98 m
Daniel	7.56 m
Edward	11.54 m
Charlie	15.46 m
George	18.86 m
Hudayfa	21.34 m
Issaa	24.86 m
Jaiden	14.54 m
Karamveer	13.27 m
Luke	19.65 m
Bilal	21.45 m

a Who threw the furthest distance?

b Who threw the shortest distance?

c What is the difference between the longest and shortest throws?

m

198

Name ....

The Sports Day finishes with the sprint races. The results for Year 5 are in the table below:

Child	Time	Child	Time
Tom	12.6 s	Bilal	11.9 s
Ben	15.0 s	Naynaa	17.6 s
Callum	11.3 s	Lola	12.7 s
Daniel	20.3 s	Ayesha	14.8 s
Edward	19.4 s	Rachel	13.6 s
Charlie	11.7 s	Ebony	18.4 s
George	14.8 s	Teghan	17.1 s
Hudayfa	16.3 s	Ursula	15.2 s
Issaa	16.0 s	Sadia	15.4 s
Jaiden	12.3 s	Wanita	17.0 s
Karamveer	13.8 s	Yvonne	16.0 s
Luke	16.7 s	Zahra	14.3 s

24 Enter the children's points for the sprint into the table on page 194.

25 Three children compare their times. The total of their times was **39.0 s**.

Which **three** children could be comparing their times?

	and		and	

26 The children find the total for the three events to find their Triathlon awards.

GOLD	SILVER	BRONZE	LEVEL 2	LEVEL 1
15 points	12 points	9 points	6 points	3 points

a Complete the table on page 194 by entering the Triathlon awards. Write **G** (GOLD), **S** (SILVER), **B** (BRONZE), **L2** (LEVEL 2) or **L1** (LEVEL 1).

b Now enter the number of awards into the table below.

	BOYS	GIRLS
GOLD		
SILVER		
BRONZE		
LEVEL 2		
LEVEL 1		

Name ....

27 **a** Enter the number of awards and stars for each house.

Each award is worth the number of stars shown below.

GOLD = 5 stars   **SILVER = 4 stars**   **BRONZE = 3 stars**

LEVEL 2 = 2 stars   **LEVEL 1 = 1 star**

	Livingstone		Nightingale		Scott	
	awards	stars	awards	stars	awards	stars
GOLD						
SILVER						
BRONZE						
LEVEL 2						
LEVEL 1						
TOTALS						

The house with the most stars wins the House Cup!

b Which house wins the House Cup?

Year 5: NUMBER - Number and place value

Page 1: 1) two hundred and seventy 2) 852 3) £1299 4) 18,062 5) ten thousand, two hundred and eighty seven 6) £22,865 7) five hundred and sixty seven thousand, eight hundred and seventy four 8) one million, six hundred and seventy three thousand, two hundred and two pounds

Page 2: 1) seventy eight 2) 235 3) eight hundred and ninety three 4) 2524 5) seven thousand, eight hundred and thirty six pounds 6) two hundred and thirty thousand, nine hundred and eighty seven 7) no; appropriate explanation 8) 1,163,457

Page 3: 1) 3652, 4652, 5652, 9652 2) 22,392 3) 352,298, 352,267, 352,200 4) 632,224 5) 736,982 6) 999,998 7) 989,000, 998,000, 1,000,009, 1,900,000 8) appropriate answer

Page 4: 1) 1782 2) 3205, 3010, 3005, 3000 3) 5436 4) 22,893, 22,793, 22,532, 22,212 5) 432,353 6) 642,837, 642,873, 738,299, 783,922 7) 1,982,998

Page 5: 1) 2 2) 3 3) 7 4) 4 5) 6 tens 6) 8 in £78,922; appropriate explanation 7) 7 in 967,899 8) 9 hundred thousands

Page 6: 1) 8 2) 7 hundreds 3) 8 thousands 4) no; appropriate explanation 5) no; appropriate explanation 6) 9 thousands 7) 9 in 891,567; appropriate explanation 8) 5 hundred thousands

Page 7: 1) 178 2) 242 3) 3342 4) 656 5) 5782 6) no; appropriate explanation 7) 74,246 8) 73,953

Page 8: 1) 22,900 2) 35,232 3) £25,600 4) 300,800 5) 773,892 6) 83,892 7) 748,946 8) 458,651

Page 9: 1) 1830 2) 4982 3) 8523 4) £6843 5) 11,753 6) 13,576 7) 4 8) 78,142

Page 10: 1) 12,320 2) 172,755 3) 326,342 4) 254,897 5) 225,900 6) 57,983 7) £673,920 8) 343,892

Page 11: 1) 8421 2) 62,800 3) £10,874 4) 21,753 5) 88,345 6) 81,847 7) 5 8) 687,632

Page 12: 1) 4892 2) 5855 3) 18,875 4) 252,834 5) 58,342 6) 81,947 7) 6 8) 798,494

Page 13: 1) -5°C 2) -1°C 3) -18°C 4) -12 5) 9 6) -14°C 7) no; appropriate explanation 8) -6

Page 14: 1) -1 2) -6 3) 3 4) -3°C 5) -14 6) -22°C 7) -£6 8) -24

Page 15: 1) 1900 ml 2) 2400 3) 12,120 4) £5300 5) 10,420 6) £12,400 7) 452,800 8) no; appropriate explanation

Page 16: 1) 100 2) 90 3) 2000 m 4) 1400 5) 82,400 6) no; appropriate explanation 7) 1,394,510 8) 9100

Page 17: 1) 2000 2) 6000 3) £13,000 4) 50,000 5) 400,000 6) 760,000 7) 1,000,000 8) £120,000

Page 18: 1) 9000 2) 83,000 3) 10,000 cm 4) 7000 5) 200,000 6) no; appropriate explanation
7) 780,000 8) 1,400,000

Page 19: 1) 12 2) 64 AD 3) 80 4) 280 miles 5) 900 BC 6) 753 BC 7) 770
8) appropriate answer

Page 20: 1) 30 2) 73 miles 3) 44 BC 4) 252 BC 5) 405 BC 6) 114 miles 7) 413 8) appropriate
answer

Page 21: 1) 620,521 2) 6 tens 3) -4°C 4) 306 AD 5) 39,828 6) 870,000 7) one million, eight
hundred and seventy three thousand, two hundred and thirty three 8) 3°C

Page 22: 1) 3 ones 2) -4 3) 87,482 4) 894,000 5) 80,000 6) 643,208 7) -3°C 8) one million,
three hundred and seventy two thousand, four hundred and ninety five

Page 23: 1) 2036 2) 14,892 3) 9 hundreds 4) 72 AD 5) -4 6) no; appropriate explanation
7) eight hundred and thirty two thousand, three hundred and ninety nine
8) 1,543,234

Page 24: 1) 30,000 2) 9 hundred thousands 3) 37,624 4) 3600 5) 4,765,730 6) 9253, 9393,
9753, 9835 7) 2 hundred thousands 8) 4,765,212

MASTERING - Number and place value

Page 25: 1) 654,465, 645,465, 546,465, 465,465, 456,465 2) 300 m 3) 13

Page 26: 4) a) no; appropriate explanation b) 36,927 5) a) MMXVII (2017) b) appropriate
answer c) IX or X

Page 27: 6) a) 800,888, 808,800, 880,088, 880,808, 880,880 b) 800,888; need to look at the
tens of thousands and thousands digits and both of these are 0 7) appropriate
6-digit numbers a) b) c) appropriate answers

Page 28: 8) 999 9) 98,764 - 10,235 = 88,529 10) a) 16:00 and 18:00 b) 10:00 and 12:00
c) winter; the temperature was very low

Page 29: 11) -50, 200 12) circled -2.5 and 0.5 13) I, IV, V, VI, IX, X, XI, XIV, XV, XVI 14) 1888
15) 61,213

Page 30: 16) a) 4 b) 5 c) 2 d) 12 floors e) 4 floors f) 2 floors g) 8 floors

Page 31: 17) a) 34,704 b) 34,740 c) 34,704 18) a) any 4 of 56,300, 56,030, 56,003, 53,600,
53,060, 53,006, 50,630, 50,603, 50,360, 50,306, 50,063, 50,036 c) appropriate
answers 19) appropriate answers

Page 32: 20) 7° 21) 46,510, 46,500, 47,000, 50,000; 23,040, 23,000, 23,000, 20,000; 84,640,
84,600, 85,000, 80,000; 181,320, 181,300, 181,000, 180,000; 65,070, 65,100, 65,000,
70,000

Page 33: 22) a) 1101, 100,011, 1210 b) 300,000 c) 3

Page 34: 23) a) 12,000, 68,000, 100,000 b) 345, 235, 435 24) lane 2; 45-50 m, lane 3; 60-70 m,
lane 4; 25-35 m

Year 5: NUMBER - Addition and subtraction

Page 35: 1) 1897 2) 4080 3) 3585 4) 6021 5) 9361 6) 24,505 7) 27,212 8) 81,128

Page 36: 1) 2991 2) 4621 3) 2415 4) 3230 5) 8251 6) 13,248 7) 31,131 8) 593,542

Page 37: 1) 2788 2) 3999 3) 6981 4) 10,000 5) 5222 6) 13,420 7) 49,550 8) 152,253

Page 38: 1) 1121 2) 299 3) 1184 4) 884; appropriate calculation 5) 1324 6) 1058 7) 52,855
8) 72,783

Page 39: 1) 1221 2) 2119 3) 2409 4) 1889 5) 12,184 6) 1886 7) 173,575 8) 454,107

Page 40: 1) 1111 2) 1326; appropriate calculation 3) 980 4) 4189 5) 2879 6) 10,866 7) 103,876
8) appropriate explanation

Page 41: 1) 7879 2) 3115 3) 6000; appropriate calculation 4) 31,188 5) 21,444 6) 60,125
7) 79,689 8) 336,965

Page 42: 1) 8687 2) 1128 3) 47,434 4) £2389 5) 216,754 6) 1,001,135
7) 729,103 – 244,647 = 484,456 or 729,103 – 484,456 = 244,647 8) 392,670

Page 43: 1) 1113 2) 1119 3) 7810; appropriate calculation 4) 11,878 5) 36,110 6) 88,881
7) 391,234 8) 6,732,766

Page 44: 1) 150 2) 201 3) 350 4) 915 5) 865 6) no; appropriate explanation 7) 4000 8) 772

Page 45: 1) 50 2) 141 3) 36 4) 102 5) 322 6) 1250 7) 2001 8) 1505

Page 46: 1) 200 2) 110 3) 279 4) 999 5) 423 6) 9100 7) 6917 8) 1268; appropriate calculation

Page 47: 1) 200 2) 60 3) 141 4) 480 5) 528 6) 1002 7) 2179 8) 5829

Page 48: 1) 130 2) 600 3) no; appropriate explanation 4) no; appropriate explanation 5) 1900
+ 2200 = 4100 6) 2300 – 1200 = 1100 7) yes; appropriate explanation

Page 49: 1) 40 + 50 = 90 2) 330 3) £170 4) no; appropriate explanation 5) no; appropriate
calculation 6) 3196

Page 50: 1) yes; appropriate explanation 2) no; appropriate explanation 3) no; appropriate
explanation 4) yes; appropriate explanation 5) no; appropriate explanation

Page 51: 1) £9 2) £62 3) £30 4) £119 5) £3.50 6) £19.50 7) no; appropriate explanation

Page 52: 1) 217 2) 70 3) 254 4) 546 5) 69 6) £85, yes; appropriate explanation 7) 482

Year 5: NUMBER - Multiplication and division

Page 53: 1) yes; appropriate explanation 2) 12, 46, 70 3) appropriate answers 4) yes;
appropriate explanation 5) no; appropriate explanation 6) yes; appropriate
explanation 7) appropriate answers 8) 345, 546

Page 54: 1) 34 2) 35, 40, 55, 600 3) yes; appropriate answers 4) no; appropriate explanation
5) 7, 14, 21, 28, 35, 42, 49 6) no; appropriate answers 7) 36, 45, 54
8) yes; appropriate examples

Page 55: 1) 1, 2, 3, 4, 6, 12 2) 3 3) yes; appropriate explanation 4) 1, 2, 11, 22 5) 6
6) 4, 6, 2, 1 7) 1, 5, 17, 85 8) 1 and 96, 2 and 48, 3 and 32, 4 and 24, 6 and 16, 8 and 12

Page 56: 1) 1, 2, 3, 5, 6, 10, 15, 30 2) no; appropriate explanation 3) 6 and 3 4) 9 5) 5, 3
6) appropriate answers 7) 1 and 72, 2 and 36, 3 and 24, 4 and 18, 6 and 12, 8 and 9
8) 1 and 140, 2 and 70, 4 and 35, 5 and 28, 7 and 20, 10 and 14

Page 57: 1) yes; appropriate explanation 2) yes; appropriate explanation 3) 2, 3, 5, 7 4) yes;
appropriate explanation 5) 63, 56, 68 6) 1 and itself; appropriate explanation
7) 107, 113 8) appropriate explanation

Page 58: 1) 13 2) appropriate explanation 3) 2, 3 (2 x 2 x 2 x 3) 4) no; appropriate explanation
5) 3, 13 (3 x 13) 6) appropriate explanation 7) 2 x 2 x 5 8) 2 x 2 x 3 x 7

Page 59: 1) 42; appropriate method 2) 138; appropriate method 3) 128 4) 352 5) 1256
6) 3896 7) 8645

Page 60: 1) 264 2) 270 3) 408 4) £1560 5) 21,315 6) 42,750 7) £16,140

Page 61: 1) appropriate explanation 2) 165; appropriate explanation 3) 105 4) 1050
5) 63 6) 5400; appropriate explanation 7) 153; appropriate explanation
8) appropriate explanation

Page 62: 1) 120 2) 1200 3) 80; appropriate explanation 4) 9 5) 550 6) 11 x 99 7) 80 8) 1080;
appropriate explanation

Page 63: 1) 19 2) 9 3) 55 4) 48 5) 169 6) 331, 2 left over 7) £565

Page 64: 1) 5, 3 remaining 2) 5, 3 remaining 3) appropriate explanation 4) yes; one person
left; appropriate explanation 5) 220 6) 224, 1 remaining 7) 225, 8 more people

Page 65: 1) 30 2) £200 3) 100 4) £3.20 5) 6000 6) 92 7) £3.60 8) £750

Page 66: 1) 5340 2) 12,000 3) 120 4) 3000 5) £2.50 6) 43.57 7) £11,500

Page 67: 1) 4 2) 25 3) yes; appropriate explanation 4) 100 5) 8 6) appropriate explanation
7) 49, 9, 81, 121 8) no; appropriate explanation

Page 68: 1) 8 2) 125 3) yes; appropriate explanation 4) 125 5) 512 6) no; appropriate
explanation 7) yes; appropriate explanation 8) 5

Page 69: 1) 25 2) 36 3) 27 4) 3045 5) 77, 1 left 6) appropriate explanation 7) 17 m 8) 63,000

Page 70: 1) 38 2) 50 3) Mr Otu, 48 miles 4) 7 miles 5) 23, 4 left over 6) Charlie, 12 lengths
7) 25

Page 71: 1) 9 m 2) 10 cm 3) 10 m 4) £300 5) £500 6) £5500

MASTERING – Addition, subtraction, multiplication and division

Page 72: 1) a) 8145 – 5687 = 2458, 8145 – 2458 = 5687 b) 9736 – 4247 = 5489, 9736 – 5489 =
4247 c) 20624 ÷ 8 = 2578 d) 586 x 9 = 5274 2) a) £1.50 b) £2 c) £24.50

Page 73: 3) a) 6, 2, 4, 4 b) 7, 5, 9, 6, 8 4) 54 5) Bilal £239.85, Tom £185.55 6) 6 hours 7) 36 days

Page 74: 8) a) 92.5 mm b) £1 c) 1.7 mm d) £3.25 9) 99

Page 75: 10) 175,176 11) £10

Page 76: 12) £78 13) a) 15 to 20 b) £3.30 to £4.40 14) yes; appropriate examples

Page 77: 15) 4 16) left to right; 2, 1, 15, 7, 4, 11, 5, 6, 14, 13 17) a) £1.60 b) £8

Page 78: 18) 250 books 19) 72, 10, 12, 2, 3, 4 20) yes; appropriate examples

Page 79: 21) 3941, 6171, 2421, 1356 22) 2.7 + 4.9 = 7.6, 4.9 + 2.7 = 7.6, 7.6 − 4.9 = 2.7, 7.6 − 2.7 = 4.9

Page 80: 23) 1 x 432, 2 x 216, 3 x 144, 4 x 108, 6 x 72, 8 x 54, 9 x 48, 12 x 36, 18 x 24 24) yes; appropriate explanation 25) yes; appropriate explanation 26) appropriate answers

Page 81: 27) a) no; 386 is not a multiple of 4 b) any three-digit number that is a multiple of 3 28) a) 156 b) 7.8 c) 6.5 d) 6.5 (appropriate explanations)

Page 82: 29) £41,184 30) 10 x 240, 12 x 200, 20 x 120, 24 x 100 etc 31) 5 (number is 61)

Page 83: 32) 24 x 100 ÷ 4 = 600, 360 x 100 ÷ 4 = 9000, 48 x 50 ÷ 4 = 600, 3.4 x 100 ÷ 4 = 85 33) a) 1 x 16, 2 x 8, 4 x 4 b) £90

Page 84: 34) b) appropriate answers c) appropriate answers d) appropriate answers e) appropriate answers

Page 85: 35) Mum 36, Grandad 64, South Street Primary School 100 36) 47 sweets

Year 5: NUMBER - Fractions (including decimals and percentages)

Page 86: 1) Tom; appropriate explanation 2) Lola 3) Sadia 4) $\frac{2}{3}$; appropriate explanation 5) most; Tom, least; Charlie 6) $\frac{5}{6}, \frac{3}{4}, \frac{8}{12}, \frac{1}{2}$ 7) no; appropriate explanation

Page 87: 1) 13 2) improper fraction; appropriate explanation 3) $2\frac{1}{2}$ 4) $1\frac{1}{2}, \frac{1}{2}$ 5) 23 6) no; appropriate explanation 7) $\frac{28}{8}$ 8) Charlie; $\frac{11}{4}$ km, Sadia; $\frac{13}{4}$ km

Page 88: 1) 11 2) improper fraction; appropriate explanation 3) $3\frac{1}{2}$ years old 4) $5\frac{2}{3}, 5$ 5) 16 6) no; appropriate explanation 7) $2\frac{6}{8}$ 8) $\frac{47}{10}$

Page 89: 1) $\frac{3}{4}$ 2) $\frac{4}{5}$ 3) $\frac{5}{7}$ 4) $\frac{7}{8}$ 5) $\frac{9}{12}$ 6) $\frac{12}{15}$ 7) $\frac{13}{16}$ 8) $\frac{17}{22}$

Page 90: 1) $\frac{1}{4}$ 2) $\frac{2}{5}$ 3) $\frac{3}{8}$ 4) $\frac{3}{9}$ 5) $\frac{5}{12}$ 6) no; appropriate explanation 7) $\frac{6}{18}$ 8) $\frac{11}{30}$

Page 91: 1) $\frac{3}{6}$ 2) $\frac{2}{7}$ 3) $\frac{5}{9}$ 4) $\frac{5}{9}$ 5) $\frac{5}{8}$ 6) $\frac{1}{12}$ 7) $\frac{11}{18}$

Page 92: 1) $\frac{3}{4}$ 2) $\frac{5}{6}$ 3) $\frac{7}{10}$ 4) $\frac{9}{12}$ or $\frac{3}{4}$ 5) $\frac{7}{8}$ 6) $\frac{10}{10}$ 7) $\frac{11}{12}$ 8) $\frac{11}{12}$

Page 93: 1) $\frac{1}{4}$ 2) $\frac{1}{6}$ 3) $\frac{4}{8}$ 4) $\frac{2}{10}$ 5) $\frac{2}{12}$ 6) no; appropriate explanation 7) $\frac{1}{12}$ 8) $\frac{2}{12}$

Page 94: 1) $\frac{4}{4}$ 2) $\frac{3}{10}$ 3) $\frac{7}{6}$ 4) $\frac{1}{1}$ 5) $\frac{6}{8}$ 6) $\frac{1}{18}$ 7) $\frac{3}{12}$ 8) $\frac{27}{18}$

Page 95: 1) 1 2) 2 3) $2\frac{1}{2}$ 4) $\frac{7}{2}$, or $3\frac{1}{2}$; appropriate explanation 5) 10 6) $62\frac{4}{6}$ 7) $3\frac{3}{4}$

Page 96: 1) ½ 2) 0.25 3) $\frac{3}{10}$ 4) yes; appropriate explanation 5) yes; appropriate explanation 6) $\frac{3}{4}$ 7) $\frac{1}{4}$ 8) $\frac{12}{10}$

Page 97: 1) $\frac{123}{1000}$ 2) $\frac{431}{1000}$ 3) 0.075 4) $\frac{2}{10}$ 5) $\frac{98}{100}$ 6) $\frac{363}{1000}$, 0.363 7) $\frac{32}{100}$ 8) yes; appropriate explanation

Page 98: 1) 2 2) 5 3) 6.3 4) 42 5) no; appropriate explanation 6) 143 7) £22 8) £78.70

Page 99: 1) 0.35 2) choc ice, toffee bar, chocolate bar 3) Ayesha ; appropriate explanation 4) 4.653 m 5) 3.089 m, 3.56 m, 3.562 m 6) 43.62, 43.671, 43.72, 43.764 7) 0.924 8) 59.664; appropriate explanation

Page 100: 1) 3.1 2) 30.81 3) 6.123 4) 9.998 km 5) 12.688 km 6) 9.302 7) 10.123 8) 15.244%

Page 101: 1) £60 2) 80 cm 3) 104 4) 75% 5) The Brainy Boffins; appropriate method 6) 25%
7) 120 8) yes; appropriate explanation

Page 102: 1) $\frac{87}{100}$ 2) $\frac{53}{100}$ 3) $\frac{67}{100}$ 4) 0.75 5) 0.52 6) $\frac{69}{100}$ 7) 31%, 0.31 8) 0.895

Page 103: 1) 0.5 2) 25% 3) 25% 4) 0.2 5) no; appropriate explanation 6) 0.1 7) 12.5% 8) 0.52

Page 104: 1) divide by 6 2) 96 ÷ 8 = 3) 72 4) 129 5) 22 6) 120 7) £660

Page 105: 1) 4 2) 24 3) 60 4) £24 5) 10 ml 6) $\frac{3}{5}$ of £70 7) $\frac{7}{12}$ 8) 16

MASTERING - Fractions (including decimals and percentages)

Page 106: 1) a) >, =, < b) >, =, < c) <, >, = 2) lines drawn correctly 3) fractions labelled correctly

Page 107: 4) a) Shop B b) 25p 5) no; $\frac{3}{8}$ of the same size shape is smaller than $\frac{3}{4}$ 6) appropriate
answers 7) appropriate answer

Page 108: 8) a) $\frac{21}{24}$; because $\frac{5}{6} = \frac{20}{24}$ b) appropriate answer 9) a) 80 b) 80 c) both answers are the
same 10) $\frac{3}{8}$

Page 109: 11) diagrams drawn correctly 12) hundredths; black 10, grey 40, white 50, tenths;
black 1, grey 4, white 5, decimal; black 0.1, grey 0.4, white 0.5, percentage; black 10%,
grey 40%, white 50%

Page 110: 13) $\frac{2}{6} + \frac{3}{9}$ 14) a) 4.05, 4.5, 5.04, 5.4, 5.50 b) 6.7, 7.07, 7.6, 7.67, 7.7 15) $\frac{5}{2} + \frac{4}{3}$ 16) 35% of
800 circled; appropriate explanation

Page 111: 17) a) 6 b) $\frac{2}{5}$ of a bottle 18) a) no; you don't know how much money each girl had to
start with b) appropriate answers

Page 112: 19) $\frac{1}{100}, \frac{2}{100}, \frac{3}{100}, \frac{4}{100}, \frac{5}{100}, \frac{6}{100}, \frac{7}{100}, \frac{8}{100}, \frac{9}{100}$ 20) February; 37,500, March; 6%, April; 52,500,
May; 8%, June; 67,500, July; 22,500, August; 4%, September; 12%, October; 82,500,
November; 112,500, December; 75,000, 10%

Page 113: 21) 0.07, 1.203 22) a) 3.455 b) between 0.761 and 0.769 23) a) $\frac{4}{20} = \frac{1}{5}$ b) $\frac{1}{15}$

Page 114: 24) circled 100 m, 400 m, 900 m, 0.75 m, 80p 25) a) $\frac{1}{4}$; b) 0.25 26) appropriate answers

Page 115: 27) $\frac{4}{20}, \frac{3}{10}, \frac{2}{5}$ 28) electric cars - 5, disabled drivers - 10, parent and child - 40,
other cars - 695 29) a) 0.24 b) 0.035

Page 116: 30) >, > 31) any from 13.35 to 13.44, 13.42 32) 36%

Page 117: 33) 48% 34) a) $\frac{18}{20}$ circled; others are all equivalent to $\frac{3}{5}$ b) $\frac{3}{9}$ circled; others are all
equivalent to $\frac{3}{10}$

Page 118: 35) 80% 36) 55% 37) 60% circled 38) $\frac{71}{100}, \frac{81}{100}, \frac{91}{100}, \frac{101}{100}$, 0.71, 0.81, 0.91, 1.01 39) 0.84, 82%,
$\frac{8}{10}$, 0.084, $\frac{8}{100}$; appropriate explanation

Page 119: 40) Lola £16, Charlie £40 41) $\frac{27}{20}$ of 60 is 81 42) 4, 30, 6 43) 0.461 to 0.489 44) 4567

Page 120: 45) $\frac{2}{5}$ circled; appropriate explanation 46) chocolate chips 45 g, rice pops 30 g, sugar
105 g 47) 24 cm

Page 121: 48) 60 49) $\frac{9}{100}$ circled 50) 9.3 and 4.9 circled 51) $\frac{11}{80}$ 52) 0.506 circled 53) 2.134 and
2.298 circled

Page 122: 54) 6, 7, 9 55) 0.067 56) peel £1.08, banana £1.80 57) 1.03 circled 58) $\frac{9}{27}, \frac{3}{9}$

Page 123: 59) $\frac{1}{2}, \frac{7}{12}, \frac{5}{8}, \frac{2}{3}$ 60) Sadia; 30 out of 50 is only 60% 61) $\frac{75}{100}$ and $\frac{3}{4}$ circled 62) 0.2 circled

Page 124: 63) £16 64) butter 168 g, flour 96 g, sugar 120 g, eggs 72 g, cocoa powder 24 g

Year 5: MEASUREMENT

Page 125: 1) 120 cm 2) Bilal; appropriate explanation 3) 32 cm 4) 1.18 m 5) 10,000 cm
6) 410 cm 7) 1.38 m 8) 4500 cm

Page 126: 1) 3600 m 2) 9.6 km 3) 4260 m 4) 6.2 km 5) 7380 m 6) Fergus; appropriate
explanation 7) 6.602 km 8) Sadia; appropriate explanation

Page 127: 1) 30 mm 2) 10 cm 3) 5 mm 4) 120 mm 5) 142 mm 6) 6.6 cm 7) 10.5 cm 8) 34.2 cm

Page 128: 1) 2 kg 2) 2300 g 3) 3250 g 4) 2.43 kg 5) 3000 g 6) Bilal; appropriate explanation 7)
15,430 g 8) 700 g

Page 129: 1) 2000 ml 2) 4 litres 3) 5.5 litres 4) 8200 ml 5) 153,000 ml 6) Fish Tank B; appropriate
explanation 7) Dad, 300 ml 8) 2000 ml

Page 130: 1) 1200 g 2) 2400 m 3) 24,500 g 4) 450 ml 5) 11.5 cm 6) 3.5 km 7) 7 litres 8) 1.8 kg

Page 131: 1) 3000 ml 2) 3500 m 3) 5 cm 4) 2.33 kg 5) 1.43 m 6) no; appropriate explanation
7) 2420 g 8) 0.47 kg

Page 132: 1) 2.5 cm 2) appropriate answers 3) 1 inch 4) 5 cm 5) 10 cm 6) 25 cm 7) 6 inches
8) 30 cm

Page 133: 1) 2.2 lbs 2) 1 kg 3) 26.4 lbs 4) 3 kg 5) 3 kg 6) Sadia; appropriate explanation
7) 15.4 lbs 8) 198 lbs

Page 134: 1) 1.75 pints 2) 3.5 pints 3) 5.25 pints 4) yes; appropriate explanation 5) Sadia;
appropriate explanation 6) 4 litres 7) 8 litres 8) 1750 pints

Page 135: 1) 18 cm 2) 15 cm 3) 48 cm 4) 24 m 5) yes; appropriate explanation 6) appropriate
explanation 7) 46 cm

Page 136: 1) 6 cm² 2) 4 cm² 3) 200 cm² 4) 4 m 5) 6 cm 6) 36 cm² 7) 96 m² 8) 51 cm²

Page 137: 1) afternoon; appropriate explanation 2) 3.30 pm 3) 17:00 4) 1.57 pm 5) 15:00
6) 15:11 7) 195 minutes 8) 150 minutes

Page 138: 1) 60 minutes 2) 210 minutes 3) 120 seconds 4) 21 days 5) 17 hours 6) 330 minutes
7) 1440 minutes 8) 4200 seconds

Page 139: 1) 70p 2) 1150 ml 3) 5800 ml 4) 120 days 5) 18 km 6) 25 7) 1.65 kg 8) 6330 g

Page 140: 1) 26 2) 9 litres 3) 2 litres 4) 2.8 km 5) 5 m 6) 1215 m 7) 1 kg 8) 10,200 g

MASTERING - Measurement

Page 141: 1) 0.7 litres, 160 cm, 1 m, 500 g 2) fence 3, fence 1, fence 2

Page 142: 3) 40 cm 4) 78 cm

Page 143: 5) a) rectangle = 5 square = 4 b) 130 cm c) 12 6) 5.20 pm

Page 144: 7) no; appropriate explanation 8) length =3 cm height = 3 cm width = 6 cm

Page 145: 9) a) smallest 16 m, largest 34 m b) £147.50 c) £2.50 10) yes; 6 x 6 = 36

Page 146: 11) a) 64 cm² b) 96 cm c) 8 cm x 8 cm x 6 cm d) cuboid e) 8 cm x 8 cm x 2 cm

Page 147: 12) appropriate explanation; pour in one full bucket, fill the bucket again and pour it into the jug, then pour the remainder of the bucket into the bowl 13) a) 3.5 m b) 77 black, 154 white c) white = 6, black = 3

Page 148: 14) 75 cm 15) a) 4 b) 8 16) length 2 cm, height 4 cm, depth 6 cm

Page 149: 17) 28 cm 18) Tom; appropriate explanation 19) 20 cm

Page 150: 20) 240 m 21) 25.5 cm

Page 151: 22) Tom 1300 m, Bilal 650 m 23) 10 cm 24) 90 cm

Page 152: 25) a) 20 cm b) appropriate drawing 26) 50 minutes

Page 153: 2) a) Bilal 33 kg, Charlie 11 kg b) £20 c) Bilal could put 11 kg of his luggage into Charlie's case 28) 20 m x 80 m

Page 154: 29) 3.75 cm 30) a) 400 b) £50,000 31) 68 cm

Year 5: GEOMETRY - Properties of shapes / Position and direction

Page 155: 1) no; appropriate explanation 2) no; appropriate explanation 3) right angle 4) no; appropriate explanation 5) 44° 6) reflex 7) angles at a point measure 360° 8) 180°

Page 156: 1) protractor 2) acute 3) no; appropriate explanation 4) right angle 5) reflex 6) 360° 7) 180° 8) obtuse, acute, acute, triangle

Page 157: 1) they are parallel 2) 4 3) 6 cm 4) 90°; appropriate explanation 5) 1 m x 15 m or 3 m x 5 m 6) 10 cm 7) 8 cm, 7 cm, 7 cm 8) 5 cm

Page 158: 1) appropriate description 2) regular; appropriate explanation 3) regular 4) irregular 5) no; appropriate explanation 6) yes 7) no; appropriate explanation

Page 159: 1) reflection 2) translate; appropriate explanation 3) reflected 4) yes; appropriate explanation 5) translate 6) no; appropriate explanation 7) appropriate explanation

Page 160: 1) square-based pyramid 2) appropriate drawing 3) appropriate explanation 4) appropriate drawing 5) 6 6) appropriate explanation 7) rectangular-based pyramid, triangular prism 8) appropriate answer

Page 161: 1) appropriate drawing 2) appropriate drawing 3) 2 4) trapezium 5) appropriate explanation 6) appropriate explanation 7) 4 8) cube

Page 162: 1) rhombus, square, rectangle, trapezium, kite 2) appropriate answers
3) yes; appropriate explanation 4) no; appropriate explanation
5) appropriate drawing 6) 4 faces, 6 edges, 4 vertices 7) appropriate description
8) appropriate investigation

MASTERING - Geometry

Page 163: 1) a) south east b) 135° anti-clockwise or 225° clockwise c) 135° 2) A will fold to make a cuboid; all the faces on A are rectangles and will fold to make a cuboid, all the faces on B are squares so will fold to make a cube

Page 164: 3) a square is a special rectangle, so if the shape had 4 right angles and 4 equal sides it could be both 4) 48° and 84° or 66° and 66°; isosceles triangle must have two equal angles 5) middle shape ticked

Page 165: 6) a) yes b) 6 equal sides, 6 equal angles 7) (2, 6) (5, 6) (5, 10) (2, 10)
OR (2, 2) (5, 2) (5, 6) (2, 6) coordinates for each solution can be listed in any order

Page 166: 8) 6th face added correctly 9) (7, 3) (10, 3) (7, 1) (10, 1)

Page 167: 10) a) A and E b) B and F c) D and C

Page 168: 11) triangles drawn correctly 12) A = 30°, B = 55°, C = 125°

Page 169: 13) a) isosceles b) 2; angles marked correctly c) BC d) AC e) a = 25° b = 90° c = 90°
d = 65° e = 25°

Page 170: 14) a) isosceles b) 45° c) 12 15) a) equilateral; the 3rd angle must be 60° b) isosceles; the 3rd angle is 55° c) scalene; the 3rd angle is 30° d) right-angled; the 3rd angle is 90°
16) a = 70°, b = 35°, c = 145°, d = 35°, e = 145°

Year 5: STATISTICS

Page 171: 1) 70 km 2) 25 km 3) lunch 4) 1.30 pm 5) 11 am – 12 noon 6) in the range of 40-50 minutes

Page 172: 1) 50 miles 2) 2 hours 3) 2.30 pm 4) 150 miles 5) 1 hour 6) 1 pm – 2 pm; lunch

Page 173: 1) 15 minutes 2) 12:15 3) 1 hour 10 minutes 4) 45 minutes 5) 3 6) 14:05 7) Train B

Page 174: 1) Train C 2) once 3) 9:55 4) 20 minutes 5) Train B 6) 25 minutes 7) 30 minutes

Page 175: 1) 1 hour 2) Bus D 3) 2 4) 10:05 5) 25 minutes 6) 15 minutes 7) 45 minutes

Page 176: 1) English and Maths 2) 10:45 3) 1 hour 30 minutes 4) Wednesday 5) 1 hour
6) 1 hour 5 minutes 7) 6 hours 15 minutes

Page 177: 1) tally or pie chart; appropriate explanation 2) line graph 3) no; appropriate explanation 4) line graph; appropriate explanation 5) no; appropriate explanation

MASTERING - Statistics

Page 178: 1) a) 2 b) some trains don't stop at every station c) 6 2) train 2; 0843

Page 179: 3) 0923 4) a) 1 hour 25 minutes b) train 5; 0943 c) train 1 and train 4 d) appropriate answer; take different routes or wait for longer at busy stations

Page 180: 5) a) 0837 b) 30 minutes c) 10:20

Page 181: 7) Monday and Tuesday 8) Monday 9) Wednesday

Page 182: 10) 3 11) Monday and Wednesday 12) Thursday; he sent more texts and emails than on any other day 13) yes; if he sent the same number he would send 140 messages every 4 weeks

Page 183: 14) a) Monday b) emails c) 6 15) 7 16) 7 text messages, 64 emails

Page 184: 17) a) every two hours b) at least 6 hours; don't know what it was at 14:00 c) it was raised and it remained the same

Page 185: d) 13:00 and 15:00 e) her temperature had returned to normal f) 15:00 g) 1.5°C h) about 37.25 °C; halfway between the two readings taken i) readings plotted correctly on graph j) the temperature scale could go up in smaller units

Page 186: 18) a) pie charts labelled correctly b) no; 24 in Year 5, 12 in Year 6 c) pie charts labelled correctly

Page 187: d) 10 e) no; 8 children in Y5 chose orange and 6 children in Y6 chose orange f) pie chart drawn and labelled correctly

INVESTIGATION - Sports Day

Page 188: 1) 96 2) 32 3) 3 4) 8

Page 189: 5) Livingstone; boys 6, girls 2, Nightingale; boys 5, girls 3, Scott; boys 2, girls 6, Total; boys 13, girls 11

Page 190: 6) 8 km 7) 20 minutes 8) 10 minutes 9) 10 minutes

Page 191: 10) 2 hours 54 minutes 11) long jump; Year 3 12:30, Year 4 12:45, Year 5 13:05, Year 6 13:30, ball throw; Year 3 13:55, Year 4 14:07, Year 5 14:22, Year 6 14:37, sprint; Year 3 14:54, Year 4 14:59, Year 5 15:06, Year 6 15:15

Page 193: 12) Luke 1.37 m, Edward 2.27 m, Callum 2.39 m, Issaa 3.17 m, Daniel 3.48 m, Bilal 3.71 m, Karamveer 3.79 m, Ben 3.97 m, Hudayfa 4.02 m, Charlie 4.06 m, Jaiden 4.56 m, Tom 4.65 m, George 5.11 m 13) 10

Page 194: 14) Tom 4 points, Ben 2 points, Callum 1 point, Daniel 2 points, Edward 1 point, Charlie 3 points, George 5 points, Hudayfa 3 points, Issaa 2 points, Jaiden 3 points, Karamveer 2 points, Luke 1 point, Bilal 2 points

Page 195: 15 & 16) Naynaa 3 m, Lola 5 m, Ayesha 2 m, Rachel 4 m, Ebony 3 m, Teghan 2 m, Ursula 4 m, Sadia 4 m, Wanita 5 m, Yvonne 3 m, Zahra 4 m 17) Naynaa 2 points, Lola 5 points, Ayesha 2 points, Rachel 3 points, Ebony 2 points, Teghan 1 point, Ursula 3 points, Sadia 3 points, Wanita 4 points, Yvonne 2 points, Zahra 2 points

Page 196: 18) even chance; same number of odd and even possibilities

Page 197: 19) $\frac{4}{11}$ 20) $\frac{2}{11}$ 21) equilateral; all angles are 60° 22) 6.19 m Teghan, 7.46 m Rachel, 10.06 m Zahra, 12.16 m Ayesha, 13.29 m Yvonne, 13.84 m Sadia, 14.92 m Ebony, 16.24 m Naynaa, 16.43 m Ursula, 18.03 m Wanita, 19.76 m Lola

Page 198: 23) Tom 4 points, Ben 3 points, Callum 5 points, Daniel 1 point, Edward 2 points, Charlie 2 points, George 3 points, Hudayfa 3 points, Issaa 4 points, Jaiden 2 points, Karamveer 2 points, Luke 3 points, Bilal 3 points, Naynaa 4 points, Lola 5 points, Ayesha 3 points, Rachel 2 points, Ebony 3 points, Teghan 1 point, Ursula 4 points, Sadia 3 points, Wanita 5 points, Yvonne 3 points, Zahra 2 points a) Callum 27.98 m b) Teghan 6.19 m c) 21.79 m

Page 199: 24) Tom 4 points, Ben 3 points, Callum 5 points, Daniel 1 point, Edward1 point, Charlie 5 points, George 3 points, Hudayfa 2 points, Issaa 2 points, Jaiden 5 points, Karamveer 3 points, Luke 2 points, Bilal 5 points, Naynaa 1 point, Lola 5 points, Ayesha 3 points, Rachel 4 points, Ebony 1 point, Teghan 2 points, Ursula 3 points, Sadia 2 points, Wanita 2 points, Yvonne 2 points, Zahra 3 points 25) Callum, Charlie, Issaa or Callum, Charlie, Yvonne or Callum, Jaiden, Sadia (accept others that total 39 seconds)

Page 200: 26) a) table completed correctly b) GOLD; boys 0, girls 1, SILVER; boys 1, girls 0, BRONZE; boys 5, girls 3, LEVEL 2; boys 5, girls 6, LEVEL 1; boys 2, girls 1

Page 201: 27) a) GOLD; Livingstone 0, Nightingale 0, Scott 1 (5 stars), SILVER; Livingstone 1 (4 stars), Nightingale 0, Scott 0, BRONZE; Livingstone 2 (6 stars), Nightingale 4 (12 stars), Scott 2 (6 stars), LEVEL 2; Livingstone 2 (4 stars), Nightingale 4 (8 stars), Scott 5 (10 stars), LEVEL 1; Livingstone 3 (3 stars) Nightingale 0, Scott 0 b) Scott